Emerging Perspectives on Gesture and Embodiment in Mathematics

A volume in
International Perspectives on Mathematics Education—Cognition, Equity, & Society
Bharath Sriraman and Lyn English, *Series Editors*

International Perspectives on Mathematics Education— Cognition, Equity, & Society

Bharath Sriraman and Lyn English
Series Editors

Mathematics Teacher Education in the Public Interest: Equity and Social Justice (2013)
Edited by Laura J. Jacobsen, Jean Mistele, and Bharath Sriraman

International Perspectives on Gender and Mathematics Education (2010)
Edited by Helen J. Forgasz, Joanne Rossi Becker, Kyeonghwa Lee, and Olof Steinthorsdottir

Unpacking Pedagogy: New Perspectives for Mathematics (2010)
Edited by Margaret Walshaw

Mathematical Representation at the Interface of Body and Culture (2009)
Edited by Wolff-Michael Roth

Challenging Perspectives on Mathematics Classroom Communication (2006)
Edited by Anna Chronaki and Iben Maj Christiansen

Mathematics Education within the Postmodern (2004)
Edited by Margaret Walshaw

Researching Mathematics Classrooms: A Critical Examination of Methodology (2002)
Edited by Simon Goodchild and Lyn English

Emerging Perspectives on Gesture and Embodiment in Mathematics

edited by

Laurie D. Edwards
Saint Mary's College of California

Francesca Ferrara
Università di Torino, Italy

Deborah Moore-Russo
University at Buffalo, State University of New York

IAP
INFORMATION AGE PUBLISHING, INC.
Charlotte, NC • www.infoagepub.com

Library of Congress Cataloging-in-Publication Data

Emerging perspectives on gesture and embodiment in mathematics / edited by Laurie D. Edwards, Saint Mary's College of California, Francesca Ferrara, Università di Torino (Italy), Deborah Moore-Russo, the University at Buffalo, State University of New York.
pages cm. – (International perspectives on mathematics education–cognition, equity & society)
ISBN 978-1-62396-553-2 (pbk.) – ISBN 978-1-62396-554-9 (hardcover) – ISBN 978-1-62396-555-6 (ebook) 1. Gesture and mathematics. 2. Communication and education. 3. Mathematics–Study and teaching. I. Edwards, Laurie D., editor of compilation. II. Ferrara, Francesca, editor of compilation. III. Moore-Russo, Deborah, editor of compilation.
QA20.G47E44 2014
510.71–dc23

2013045517

Printed in the United States of America

CONTENTS

SECTION III

GESTURE AND EMBODIMENT IN UNIVERSITY MATHEMATICS

INTRODUCTION

Deborah Moore-Russo
University at Buffalo, State University of New York

Francesca Ferrara
Università di Torino, Italy

Laurie D. Edwards
Saint Mary's College of California

INTRODUCTION

The notion that doing, learning, and communicating about mathematics involves the body in significant ways has received attention of late from many scholars within the field of mathematics education; one form that this research has taken is attention to gesture in mathematical practices (e.g., Arzarello, Paola, Robutti, & Sabena, 2009; De Freitas & Sinclair, 2012; Edwards, 2009; Healy & Fernandes, 2011; Nemirovsky & Ferrara, 2009; Núñez, 2009; Radford, 2009, 2011; Radford, Edwards, & Arzarello, 2009; Thom & Roth, 2011). The purpose of this volume is to bring together both seminal and recent research on gesture and embodiment as it applies to mathematical thinking, in learners ranging from young children to graduate students. While acknowledging that mathematics is socially constructed, we propose that its construction is not arbitrary or unconstrained, but rather is rooted in, and shaped by, the body.

Emerging Perspectives on Gesture and Embodiment in Mathematics, pages vii–xii

The study of cognition as a whole has been reconceptualized during recent decades. In place of a focus on formal structures, abstracted away from lived experience, fields ranging from cognitive science, neuroscience, psychology, and linguistics have gathered evidence that thought is closely linked to action, constructing a second generation of theories of cognition that attend to physical action as well as perception and other neurological processes (e.g., Clark, 2001; Gallagher, 2005; Gallese & Lakoff, 2005; Varela, Thompson, & Rosch, 1991). These new approaches to the study of cognition, as well as recent developments in semiotics and the study of communication, are grounded in close observation of multiple bodily-based modalities, that is, physical or sensory resources for making meaning (Edwards & Robutti, this volume; Kress, 2001, 2005). These modalities include speech and other verbalizations, written inscriptions (including words, symbols and graphics), gesture, other bodily actions/stances, and physical interactions with objects in the world or with virtual "objects" on computer screens.

It is possible to view non-verbal modalities as potential sources of information about what is going on "in the heads" of learners, but not to see them as being involved in the generation of thoughts or ideas themselves. From this perspective, cognition, although embodied in an obvious and uncontroversial way in that it takes place in the brain, is not connected to actions such as gesture or bodily movement, or to other non-verbal modalities. In other words, although communication may take a bodily form, cognition is still seen as an abstract function of the mind.

This volume takes a different perspective. The researchers in mathematics education represented in this volume, as well as many others, believe that bodily modalities are not epiphenomenal to cognition, but rather are integral components of mathematical thinking, teaching and learning. From the perspective of embodied cognition, all who engage with mathematics, whether at an elementary or an advanced level, share basic biological and cognitive capabilities as well as common physical experiences that are inextricably linked to being human and living in a material world (Lakoff & Núñez, 2000; Núñez, 2008). It is precisely because the body is so important in our human interaction and activity that it is vital to also consider it with respect to the learning and teaching of mathematics. Hence, this book looks at the doing and communicating of mathematics as more than a purely intellectual activity, but rather one that is bodily based, involving a wide range of physical actions, including gestures.

The volume presents recent research and theory on gesture and mathematics within a framework that addresses various levels of mathematical development. The first chapter by Edwards and Robutti addresses the foundational concepts of embodiment and modality, examining in particular the affordances of different modalities for creating and expressing mathematical meanings. After this, the book is divided into three sections by level of mathematical development.

Section I considers gestures and embodiment in early mathematics. In Chapter 2, Alibali, Church, Kita, and Hostetter make a case that gestures can signal changes in children's developing embodied knowledge of conservation of quantity. In this context, gestures can be seen as an important bridge from perception- and action-based knowledge to more abstract knowledge that can be expressed in words. Goldin-Meadow, Levine and Jacobs focus, in Chapter 3, on the role that gesture plays in the teaching and learning of counting and simple arithmetic. In analyzing the gestures of both young learners and adult teachers, they demonstrate that gesture can not only provide a window on children's mathematical thought, but a means of shaping it.

The three chapters in Section II look at gestures and embodiment in school mathematics. In Chapter 4, Arzarello and Sabena highlight the relationship between speech and gestures, considering how the gesture-speech unity goes beyond semantics and contributes to the mathematical argumentation of 10-year olds. In their analysis, gestures not only complement speech in a global-holistic way, but also can provide logical structure within mathematical discourse. Ferrara, Robutti and Edwards, in Chapter 5, study how one fifth-grade student demonstrates multimodal expressivity in her gestures, speech, head movements, and voice intonation as she constructs and expresses her knowledge during a class investigation of uniform motion. In this textured examination of the interplay among modalities, the argument is made that an important aspect of mathematical learning is the bridging of students' personal embodied and sensory experiences with the communal construction of mathematics in the classroom. In Chapter 6, Healy and Fernandes focus on the gestures of blind learners as they engage in area and volume tasks. The case is made that gestures serve as a means by which these learners create and express embodied abstractions associated with geometric properties, methods and relationships. In addition, the communicative potential of gestures among the blind students, as well as between the students the sighted teachers, is explored.

Section III contains five chapters related to gestures and embodiment in university-level mathematics. In Chapter 7, Moore-Russo and Viglietti examine future and practicing mathematics teachers' gestures and use of objects while investigating three-dimensional space. They argue that in both research and teaching, we must attend to the mind and the body, in interaction with the physical environment, to fully understand and support learning. In Chapter 8, Yoon, Thomas, and Dreyfus focus on the conscious, deliberate mimicry of gesture as they follow two secondary mathematics teachers' efforts to solve calculus tasks. They posit that such mimicry can support the appropriation of mathematically meaningful gestures as well as the grounded blends that support them, enhancing learning. In Chapter 9, Sweeney and Rasmussen view modeling as an embodied organizing activity as they analyze two engineering students engaging with a double Ferris wheel, shown in an interactive computer simulation. From their perspective, modeling provides the opportunity for connecting perceptual and bodily experiences in a mathematical context rich with possibilities for recording and reflecting on actions.

In Chapter 10, Marghetis, Edwards, and Núñez examine embodiment and gesture in a context involving mathematics doctoral students formulating a proof. In a fine-grained analysis of gestures produced during the proving process, the authors find a relationship between the discourse content and the kind of gesture used by students. Finally, in Chapter 11, Smith considers the ways in which gestural and other embodied mechanisms are blended in the discourse among advanced mathematics students working together to complete homework problems. The importance of real-space blends that link conceptual and physical realities is explored, as well as the rich and changing relationships among the entire range of communicative and cognitive modalities.

We are pleased to offer this collection of work addressing embodiment, gesture, and multimodality across a range of mathematical topics and among learners of various ages. We hope that the book will make a contribution to the ongoing investigation of how human beings create and make sense of mathematical ideas.

REFERENCES

Arzarello, F., Paola, D., Robutti, O., & Sabena, C. (2009). Gestures as semiotic resources in the mathematics classroom. *Educational Studies in Mathematics, 70*(2), 97–109.

Clark, A. (2001). *Being there: Putting brain, body, and world together again.* Cambridge, MA: Bradford Press.

De Freitas, L., & Sinclair, N. (2012). Diagram, gesture, agency: Theorizing embodiment in the mathematics classroom. *Educational Studies in Mathematics, 80*(1/2), 133–152.

Edwards, L. D. (2009). Gestures and conceptual integration in mathematical talk. *Educational Studies in Mathematics, 70*(2), 127–141.

Gallagher, S. (2005): *How the body shapes the mind.* Oxford, England: Clarendon Press.

Gallese, V., & Lakoff, G. (2005). The brain's concepts: The role of the sensory-motor system in conceptual knowledge. *Cognitive Neuroscience, 22,* 455–479.

Healy, L., & Fernandes, S. H. A. A. (2011). The role of gestures in the mathematical practices of those who do not see with their eyes. *Educational Studies in Mathematics, 77,* 157–174.

Kress, G. (2001). *Multimodal discourse: The modes and media of contemporary communication.* London, England: Arnold Publishers.

Kress, G. (2005). Gains and losses: New forms of texts, knowledge, and learning. *Computers and Composition, 22,* 5–22.

Lakoff, G., & Núñez, R. (2000). *Where mathematics comes from: How the embodied mind brings mathematics into being.* New York, NY: Basic Books.

Nemirovsky, R., & Ferrara, F. (2009). Mathematical imagination and embodied cognition. *Educational Studies in Mathematics, 70*(2), 159–174.

Núñez, R. (2008). Mathematics, the ultimate challenge to embodiment: Truth and the grounding of axiomatic systems. In P. Calvo, & T. Gomila (Eds.), *Handbook of cognitive science: An embodied approach* (pp. 333–353). Amsterdam, The Netherlands: Elsevier.

Núñez, R. (2009). Gesture, abstraction, and the embodied nature of mathematics. In W.-M. Roth (Ed.), *Mathematical representation at the interface of body and culture* (pp. 309–328). Charlotte, NC: Information Age Publishing.

Radford, L. (2009). Why do gestures matter? Sensuous cognition and the palpability of mathematical meanings. *Educational Studies in Mathematics, 70*(2), 111–126.

Radford, L. (2011). Embodiment, perception and symbols in the development of early algebraic thinking. In B. Ubuz (Ed.), *Proceedings of the 35th Conference of the International Group for the Psychology of Mathematics Education* (Vol. 4, pp. 17–24). Ankara, Turkey: PME.

Radford, L., Edwards, L. D., & Arzarello, F. (2009). Introduction: Beyond words. *Educational Studies in Mathematics, 70*(2), 91–95.

Thom, J. S., & Roth, W.M. (2011). Radical embodiment and semiotics: Toward a theory of mathematics in the flesh. *Educational Studies in Mathematics, 77,* 267–284.

Varela, F., Thompson, E., & Rosch, E. (1991). *The embodied mind: Cognitive science and human experience.* Cambridge, MA: MIT Press.

CHAPTER 1

EMBODIMENT, MODALITIES, AND MATHEMATICAL AFFORDANCES

Laurie D. Edwards
Saint Mary's College of California

Ornella Robutti
Università di Torino, Italy

No ideas
but in things.
—William Carlos Williams

Mathematics is often considered to be a subject far removed from the mundane; Gregory Bateson (2000) called it a "rigorous fantasy... forever isolated by its axioms and definitions from the possibility of making an indicative statement about the real world" (p. 428). The purpose of this chapter is to consider the nature of mathematics from an embodied perspective, in which mathematical ideas are assumed to be like other human conceptions, in that they emerge from the interaction between an individual and the world and among individuals through time. An additional aim is to examine how mathematical ideas are constructed and expressed, again working from an embodied perspective. Our thesis is that, far from being a lifeless

Emerging Perspectives on Gesture and Embodiment in Mathematics, pages 1–23

abstraction, mathematics is a human cultural creation grounded in physical experience and expressed through multiple semiotic and bodily-based modalities. In this chapter, we examine the terms embodiment and multimodality as applied to mathematical thinking, and we analyze the affordances and constraints offered by different modalities in doing mathematics.

Attention to the role of the body in mathematics is consistent with the expanding scope of inquiry within the field of mathematics education, which initially investigated mathematics learning more or less at a distance, via surveys and written examinations, but later gathered data directly from individual learners (Kilpatrick, 1992). In recent decades, mathematics education experienced a "turn to the social" (Lerman, 2000), in which sociocultural factors, including interaction and discourse, were acknowledged as essential. This was followed by a nascent "turn to the body" (Lakoff & Núñez, 2000), in which mathematical ideas, like other forms of cognition, are viewed as the product of embodied human existence (Johnson, 2007; Lakoff & Johnson, 1999).

From the perspective of embodiment, although mathematics may be socially constructed, this construction is not arbitrary or unconstrained but rather is rooted in and shaped by the body (Núñez, 2008; Núñez, Edwards, & Matos, 1999). The doing and communicating of mathematics is never a purely intellectual activity; it involves a wide range of bodily actions, from committing inscriptions to paper or typing equations into a computer, to speaking, listening, gesturing, and gazing. Each of these different modalities offers a different set of potentialities to the person who is doing mathematics. Thus, the focus of this chapter is on the nature of embodiment and multimodality in mathematics, and how embodied resources can contribute to mathematical practice.

EMBODIED MATHEMATICS

The idea that mathematical knowledge arises from experience in the physical world is not new. Nearly three decades ago, the mathematician and physicist Dirk Jan Struik (1986) distinguished between the symbolic forms of mathematics and its origins in "the world of experience":

> Its abstract symbolism can blind us to the relationship it carries to the world of experience. Mathematics, born to this world, practised by members of this world with minds reflecting this world, must capture certain aspects of it—e.g., a "number," expressing correspondences between sets of different objects; or a "line," as the abstract of a rope, a particular type of edge, lane or way. The theorem you discover has not been hauled out of a chimerical world of ideas, but is a refined expression of a physical, biological, or societal property. (p. 286)

Yet to many immersed in the layers of abstractions of mathematics, it may seem to be the least likely domain to claim a connection to the physical world, much less to the human body. After all, David Hilbert famously pointed out the abstract nature of mathematics when he said, "One must be able to say at all times—instead of points, straight lines, and planes—tables, chairs, and beer mugs" (Hilbert, cited in Reid, 1996, p. 57), implying that the "objects" of mathematics are but placeholders within an essential structure of logical relationships. In asking how abstract mathematics is connected to the body, however, we may be posing the question the wrong way around. If we begin our analysis with the body instead of with the domain of mathematics, the question then becomes: What kind of mathematics can human beings create, with the kind of bodies, minds and brains that we possess?[1] In a sense, we already have the answer: It is the mathematics that we, as a species, have created (Lakoff & Núñez, 2000; Núñez, 2008). But rather than starting with the finished (cultural) product and seeking bodily roots, it may be more helpful to begin with the body and trace a possible path whereby abstractions like mathematics can emerge.

Nemirovsky (2003) states that, "mathematical abstractions grow to a large extent out of bodily activities having the potential to refer to things and events as well as to be self-referential" (p. 103). Thus, the essential starting place for understanding cognition is in its relationship to the body. An important point to remember is that both evolutionarily and developmentally, the mind has evolved to control the body, within specific and ever changing external circumstances. It did not evolve primarily in order to process symbols or engage in purely intellectual thought. Clark (2001) contrasts a view of the mind as a processor of symbols (common in "first-generation" cognitive science; Gallese & Lakoff, 2005) with a contemporary embodied view:

> In place of the intellectual engine cogitating in a realm of detailed inner models, we confront the embodied, embedded agent acting as an equal partner in adaptive responses which draw on the resources of mind, body, and world.... The idea here is that the brain should not be seen primarily as a locus of inner *descriptions* of external states of affairs; rather, it should be seen as a locus of inner *structures* that act as operators upon the world via their role in determining actions. (Clark, 2001, p. 47)

In other words, the mind is first a controller of bodily action. Whatever else it may be, it is built on that foundation. Because bodily action often involves the manipulation of objects in the world, including, importantly, the control and manipulation of one's own body, this experience constitutes a universal source domain for constructing many kinds of understandings, including mathematical ones. The significance of our experiences with objects as a source domain for mathematical understandings will be discussed next.

Physical Objects and Mathematical Objects

Our initial human experiences, both developmentally and evolutionarily, are with physical objects rather than symbols. That is, for an infant (or a human being before language was "invented"), interaction with the world does not take place through formal, conventional symbols, but through things. The emergence of language and symbols, in the individual and the species, occurs only after a sustained period of contact with objects via physical actions. Thus, the "primal" knowledge we gain when we learn to physically manipulate objects is available to all human beings (including practitioners of mathematics) as they engage with less concrete entities. Although experience with objects as a source domain in mathematics may be difficult to discern, precisely because such experience is so ubiquitous, an examination of mathematics discourse and texts reveals its influence. When a teacher tells a student to "carry the 1" or "integrate the function," he or she is implicitly assuming that the student can think about a number or function as a "thing" that can be acted on. Even as researchers, we tend to talk about mathematical "entities" or "objects," which, from the point of view of formal mathematics, is inappropriate because these "entities" are explicitly stated to be noncorporeal, and their characteristics are based solely on logical properties and definitions rather than worldly attributes. Font and his colleagues discuss the differences between physical objects, which are ostensive (i.e., to which one can point; Quine, 1950), and mathematical objects, which are not (Font, Godino, Planas, & Acevedo, 2010):

> In mathematical discourse it is considered, whether explicitly or implicitly, that mathematical objects exist in a special way (non-ostensive, virtual, ideal, mental, abstract, general, etc., depending on the theoretical perspective) that is different from the way in which physical objects exist, and which particularly differs from the material symbols that represent them . . . speaking about the existence of mathematical objects, as objects that exist in a form that is different from that of their material symbols, is essentially a metaphorical question. (p. 15)

Font et al. (2009, 2010) spell out the manner in which teachers, learners, and mathematicians talk about and "manipulate" nonostensive mathematical entities as if they were things. This occurs through a metaphor or conceptual mapping between the source domain of physical objects and the target domain of mathematical entities, a mapping that projects the inferential structure of the source domain onto the target domain (Lakoff, 1993). Thus, we are able to talk about an entire range of mathematical "entities" (e.g., sums, integrals, groups) as if they were objects with independent existence (rather than socially shared labels for mentally constructed patterns and structures).

Font et al. (2009) call this mapping the "object metaphor" or the "mathematical entities as physical objects" metaphor. Note that this is an unconscious conceptual metaphor (or a single scope blend; see Fauconnier & Turner, 2002). Undertaking what they call "mathematical idea analysis," Lakoff and Núñez (2000) have identified a wide range of such metaphors within mathematical discourse. Not only does the object metaphor show itself in speech and written text, but it is also found in gesture and imagery. When discussing various mathematical entities, we may point to locations in gesture space or delineate regions on a piece of paper, actions that would not make sense unless we were thinking of these entities as ostensive objects. The object metaphor is a foundational metaphor in mathematical thought and activity, and its pervasiveness (yet "invisibility") provides an important example of the way that physical embodiment is deeply woven into mathematics.

Thus, from the perspective of embodiment, our interactions with the world involving physical objects provide the input space or source domain for the kind of nonphysical entities we later "create" by proposing and accepting shared definitions in mathematics.

Another expression of the object metaphor can be found in the often-studied cognitive transformation of mathematical actions or processes into so-called "objects" (e.g., Dubinsky & Harel, 1992; Sfard, 1994; Tall, Thomas, Davis, Gray, & Simpson, 2000). Just as in the physical world, we might engage in an action or a process called "walking" and later talk about going for a "walk" (where "a walk" is now considered a "thing" rather than an action), in mathematics, many mathematical concepts are first experienced as processes and later referenced as objects. For example, students learn first to solve a problem (a process) but later can step back to consider whether their solution (an entity) is optimal. Algebra students first engage in plotting and connecting points on a grid and later think of the outcome of these processes as an object, a mathematical function. This is an important conceptual capability when doing mathematics; without it, we would find it difficult to abstract or generalize, to use "compressed" versions of processes as inputs in building new mathematical entities and patterns (Fauconnier & Turner, 2002). We would argue that the fundamental metaphor of "mathematical entities as physical objects" makes possible the intellectual work that allows us to transform mathematical processes into objects.

Symbols and Inscriptions

Algebra and its use of non-numerical symbols is based on a common linguistic mechanism, which Lakoff and Núñez (2000) call the Fundamental Metonymy of Algebra. Metonymy refers to the ability to use a specific part or aspect of a referent to stand for the whole, as when we say "the White House" when we actually mean the U.S. president. The Fundamental Metonymy of Algebra is defined as a "metonymic mechanism that makes the discipline of algebra possible, by allowing us to reason about numbers or other entities without knowing which particular entities we are talking about" (Lakoff & Núñez, 2000, p. 75). Utilizing this linguistic mechanism, a letter or phrase can stand for an unknown or a range of possible values (e.g., the values that make the equation "$x + y = 12$" true; the set of natural numbers). Building on the fundamental metonymy that letters can stand for numerical values, the metaphor "mathematical symbols are physical objects" makes it possible to manipulate and "move" mathematical symbols as if they were physical objects. Of course there are more constraints on the manipulation of mathematical symbols (and what they stand for) than there are on physical objects because of the formal definitional nature of mathematical entities. However, this metaphor can help us understand a relatively common type of error in learning algebra, which occurs when the definitional constraints are ignored in favor of the more flexible affordances of physical objects, leading to "symbol-pushing" (e.g., Arcavi, 1994; Kieran, 1992; Matz, 1982).

Written (or displayed) mathematical symbols are an example of external inscriptions, without which mathematical actions and results would be ephemeral and easily forgotten. We use the term "inscription" to refer to an external "representation," whether symbolic or imagistic, which is nonephemeral and therefore amenable to reflection, review, and revision. The invention of mathematical symbols is one example of the development of external representational scaffolding, which includes the range of methods humans have created to keep track of thoughts and prior actions (including language, imagery, and concrete artifacts). Yet the use of external structure to keep track of or "represent" thought is not a simple unidirectional process. That is, representations do not simply "carry" meanings from one person to another or from the mind to the external world, as if they were "conduits" of information (Edwards, 1995; Lakoff, 1993; Reddy, 1979). Instead, inscriptions are important elements in a process of feedback and feedforward, in which the act of creating external representations can change what one is trying to represent (e.g., Clark, 2001).

Mathematics is clearly a domain in which the use of external scaffolding, in the form of conventional symbols as well as both conventional and idiosyncratic graphical inscriptions, has contributed in vital ways to the evolution of the domain (Cajori, 1993). Hutchins (2005) would call these inscriptions "material anchors," and, again, from an embodied perspective, inscriptions do not simply "represent" an internal collection of preexisting thoughts. Instead, mathematical arguments and ideas are developed through the iterative practice of recording, manipulating, considering, erasing, and rewriting symbols while using the record created with symbols to test and refine the ideas themselves. Clark (2001) describes this affordance of writing in the more general case of creating "text":

> By writing down our ideas, we generate a trace in a format that opens up a range of new possibilities. We can then inspect and reinspect the same ideas [...]. We can hold the original ideas steady so that we may judge them, and safely experiment with subtle alterations. We can store them in ways that allow us to compare and combine them with other complexes of ideas in ways that would quickly defeat the unaugmented imagination. In these ways [...], the real properties of physical text transform the space of possible thoughts. (p. 208)

Clark (2001) goes on to discuss how written language and symbols make possible what he calls "second-order cognitive dynamics," the ability to think about our own thinking, including, "coming to see why we reached a particular conclusion by appreciating the logical transitions in our own thought" (pp. 208–209), an essential aspect of mathematical reasoning. Clark (2001) describes the way that viewing thoughts as objects makes possible metacognitive processes:

> As soon as we formulate a thought in words (or on paper), it becomes an object for ourselves and others. As an object, it is the kind of thing that we can have thoughts about. [...] The process of linguistic formulation thus creates the stable structure to which subsequent thinkings attach. (p. 209)

Thus, the fact that human minds are fundamentally embodied controllers of action, while we exist in a cultural world that offers scaffolding for actions and thoughts, provides a starting point in considering the particular modalities utilized in thinking and doing mathematics.

MEANINGS FOR MULTIMODALITY

The term "multimodality" has been used in many different fields and analytic contexts, ranging from the study of communication to examinations of neurological processes. From the discussion below, it should be apparent that the meanings used in these different fields of study are not mutually exclusive but intersect and complement each other. We will briefly discuss these usages and conclude by stating how we use the term in analyzing mathematical thinking.

Sensory Modalities

Aristotle is held to have distinguished five senses, and these are often considered to be the primary sensory modalities (or channels): sight, hearing, touch, taste, and smell. According to contemporary physiologists, five characteristics define a sensory modality (Kling & Riggs, 1971):

> They have (1) *markedly different receptive organs* that (2) respond to *characteristic stimuli*. Each set of receptive organs has its (3) *own nerve* that goes to a (4) *different part of the brain,* and the (5) *sensations are different.* (p. 118; italics in original)

Based on the criteria above, four additional sensory modalities have been identified in humans, beyond those specified by Aristotle: kinesthesia (joint sense), vestibular sense (balance as signaled by the inner ear), temperature sense, and pain. These four are called "somatosensory" modalities.

Clearly, sensory modalities make up an important element of learning, whether of mathematics or other subjects, for it is only via the senses that a learner has access to either direct experience or culturally transmitted knowledge. Although sight and hearing are the most important modalities used in formal schooling, other sensory modalities may provide unconscious grounding for understanding fields like mathematics. For example, the sense of balance provided by one's vestibular sense can provide the foundation for understanding the algebraic process of "balancing" an equation, among other mathematical concepts (Johnson, 1987; Núñez, Edwards, & Matos, 1999), and the experience of touch underlies the comprehension of the "behavior" of a function near an asymptote.

Neural Multimodality

Vittorio Gallese, a neuroscientist, and George Lakoff, a linguist, utilize the term "multimodality" in a specific way in their model of how concepts are created in the brain. This model offers an alternative to the information-processing stance toward cognition, in which it is held that perception, thought, and motor action are three separate brain processes (Barsalou, 2008). In the information-processing model, the perceptual system first takes in outside stimuli, which are then processed in an "association area" in the cortex. The cortex subsequently directs action through the premotor and motor cortices, resulting in a possible action in response to the stimulus. In contrast, Gallese and Lakoff (2005) propose an interactionist theory built on recent discoveries that, in addition to action-only or perception-only neurons, neuron assemblages in the premotor and parietal areas do two things at once: respond to sensory input and initiate or simulate action. One particular neuron of this kind is called a "mirror neuron," which acts in the following way:

> [M]irror neurons [...] discharge when the subject (a monkey in the classical experiments) performs various types of hand actions that are goal-related and also when the subject observes another individual performing similar kinds of actions. (Gallese & Lakoff, 2005, p. 460)

In other words, certain neurons are activated not only by particular actions but also by seeing such actions performed by others.

Gallese and Lakoff (2005) characterize this linkage of perception and action as "multimodality" at the neuronal level. They also note that the entire sensorimotor system, as well as language itself, is multimodal:

> Circuitry across brain regions links modalities, infusing each with properties of others. The sensory-motor system of the brain is thus "multimodal" rather than modular. Accordingly, language is inherently multimodal in this sense, that is, it uses many modalities linked together—sight, hearing, touch, motor actions, and so on. Language exploits the pre-existing multimodal character of the sensory-motor system. If this is true, it follows that there is no single "module" for language—and that human language makes use of mechanisms also present in nonhuman primates. (p. 456)

Based on this conception of multimodality, Gallese and Lakoff (2005) propose a redefinition of "concept," one quite different from that found in classical cognitive science. In "first-generation" cognitive science, the definition of a concept is based on a set of necessary and sufficient conditions, and concepts are seen as "modality-neutral and symbolic" (p. 466). However, according to Gallese and Lakoff, concepts are embodied: They arise as a consequence of human action or internal simulation of such action, through the formation of clusters of functional neurons within larger structures they call schemas. For Gallese and Lakoff (2005), these schemas are unlike the purely internal schemas described by Piaget or information-processing psychology:

> Schemas are *interactional*, arising from (1) the nature of our bodies, (2) the nature of our brains, and (3) the nature of our social and physical *interactions* in the world. Schemas are therefore not purely internal, nor are they purely representations of external reality. (p. 466)

This description of schemas as interactional, as well as the work on neural multimodality, offers a biologically grounded basis for a theory of embodied mathematics. Although it may not be on the immediate horizon, it is reasonable to foresee the eventual identification of neurally based schemas and concepts for specific mathematical ideas.

Multimodality in Communication

In recent decades, linguists, semioticians, and other scholars interested in discourse have drawn attention to the fact that communication occurs in ways that go beyond oral speech and written language, introducing the notion of semiotic multimodality. For example, Kress (2001b) describes multimodality as "the idea that communication and representation always draw on a multiplicity of semiotic modes of which language may be one" (pp. 67–68). Researchers in mathematics education have also fruitfully utilized a semiotic approach in the examination of the multiple means of expression found in mathematical practice, including spoken words, mathematical symbols, and various kinds of imagery, such as gesture (e.g., Arzarello, Paola, Robutti, & Sabena, 2009; Arzarello & Robutti, 2008, 2010; Radford, 2009, 2011).

Among the primary semiotic modes discussed by Kress and others (e.g., Bateman, 2009; Norris, 2004) are language, imagery, and sound. With the evolution of the discipline, semioticians now also look at more complex and broad-ranging modes, including music, theatre, color, clothing, and even furniture layout (see Kress & Van Leeuwen, 2002). From this perspective, virtually any means that humans use to express or organize themselves can be seen as a semiotic mode:

> I use the term "mode" for the culturally and socially produced resources for representation and "medium" as the term for the culturally produced means for distribution of these representations-as-meanings, that is, as messages. These technologies—those of representation, the modes and those of dissemination, the media—are always both independent of and interdependent with each other. (Kress, 2005, pp. 6–7)

Two aspects of the definitions above should be noted: Kress restricts modes (which we take as a synonym for modalities) to resources that are "culturally and socially produced." He also distinguishes between modes (for representation) and media (as methods of dissemination). This distinction is further clarified when he states:

> Media are the material resources used in the production of semiotic products and events, including both the tools and the materials used (e.g., the musical instrument and air; the chisel and the block of wood). They usually are specially produced for this purpose, not only in culture (ink, paint, cameras, computers), but also in nature (our vocal apparatus). (Kress, 2001a, p. 22)

Within this framework, Kress does not explicitly discuss the sources of the meanings or messages that are being represented or disseminated, yet his language suggests a conduit metaphor of representation (Lakoff, 1993; Reddy, 1979). In such a metaphor, an idea or a message originates (presumably in abstract form) in the subject's mind and is then expressed or transferred via one or more semiotic modes (language, imagery, sound, etc.). These modes, in turn, are made concrete via particular material media. For example, the (abstract) mode of language can be delivered through the spoken word, in writing or type on paper, via characters on a computer screen, and so on.

The metaphor of communication or representation as conduit and the conceptualization of modes as purely social or cultural resources reveal important differences with the theory of embodied cognition. According to this theory, ideas do not originate as abstractions that are made concrete through particular media. Instead, the generation of ideas and concepts is intimately linked to motor action, as well as simulated or imagined motor action, as discussed by Gallese and Lakoff (2005). Ideas are embodied from the start, based on individual experience and human physical capabilities. In addition, from an embodied perspective, modes are not restricted to "culturally and socially produced resources" (Kress, 2005, p. 6). Instead, the body offers numerous resources for creating and expressing meanings, including gesture, bodily stance and movement, gaze, rhythm, and prosody in speech.

In order to incorporate these additional means for making and expressing ideas, we propose a theoretical framework for multimodality that integrates the body and looks more closely at how concepts originate in the embodied mind.

AN EXPANDED VIEW OF MODALITIES

From the perspective of embodied cognition, bodily resources are vital in the production of meanings, not just in communicating them (Barsalou, 2008; Clark, 2001; Gallese & Lakoff, 2005; Goodwin, 2003; Johnson, 2007; Lakoff & Johnson, 1999; McNeill, 1992, 2005). This is true within mathematics no less than within any other domain. A student who painstakingly plots the points of a function for the first time and connects them into a (more or less) smooth curve is not simply expressing concepts that already exist, conveyed via the medium of pencil and paper. His or her physical engagement with the graph paper and pencil, and the iterative action of consulting a table of values, locating and plotting those values, we would argue, are essential aspects of the construction of the concept of a graph of a function. Later work with graphs may include other modalities, perhaps entering equations into a computer or the intentional production of gestures (see Gerofsky, 2010). The body is thus not simply a medium but an important resource in the construction and communication of meaning. It is also, clearly, a vital element in receiving meanings generated by others via the sensory modalities.

Thus, we propose a broader definition for modality that encompasses and goes beyond the traditional notion of semiotic mode. Within this expanded perspective, we see modalities as the entire range of cultural, social, and bodily resources available for receiving, creating, and expressing meaning. In addition to sensory modalities, which receive information, this category would also include motor modalities, such as gesture, bodily stance, touch, and so on—essentially anything that humans can do with their bodies to communicate or construct ideas. Along with Kress, we also see the body as a medium for the expression of ideas; however, we see it not simply as a medium but also as a primary modality for thought.

Our framework also includes a category for the expressive products created by humans through the use of language, imagery, bodily motion, or any other modality. By "expressive product," we refer to the physical "traces," whether permanent or ephemeral, of people's actions. These may take the form of writing and other inscriptions, utterances, song, dance, computer imagery, physical constructions, or any other observable production. Expressive products are sometimes called "representations"; however, this term has often been used to imply the existence of an abstract internal meaning that is simply mapped onto an external representational system. Because we do not adopt this perspective, we have chosen to use the term "expressive product" rather than "representation."

Table 1.1 presents an outline of this four-category framework for understanding multimodality. The categories include bodily modalities (both sensory channels and motor actions), semiotic modes, material media, and concrete expressive products.

Often we find that research in mathematics education attends to only one or perhaps two of these modalities, which is understandable given the richness of each mode of expression. However, all four categories in Table 1.1 tend to be involved in living acts of communication or cognition. If we examine the semiotic mode of language, for example, it might take the form of oral speaking. Speech, seen as both a motor action and an expressive product, is mediated via air moving through the lungs, larynx, and mouth. It requires the motor activity of these bodily parts, as well as the use of the sensory channel of hearing. Although the paradigmatic situation for speech involves a speaker and interlocutor in physical proximity, oral speech might also be transmitted via a different medium, such as a telephone or radio.

As a second case of the use of language as a semiotic mode, we can consider signed communication by the Deaf. A deaf person might not use oral speech for communication; instead, she might express herself, for

TABLE 1.1 A Framework for Multimodality

Bodily Modalities

Sensory
- Sight, hearing, touch, vestibular, etc.

Motor
- Motor actions in general
- Gesture
- Gaze
- Head movement
- Full body movement
- Bodily stance
- Manipulation of artifacts
- Prosody, rhythm, etc.

Semiotic Modes

Sensory and motor
- Language
- Mathematical symbols
- Musical notation
- Other formal notation systems
- Visual imagery (external)
- Sound
- Clothing, architecture, dance, colors, etc.
- Any cultural system for making meaning

Material Media

Bodily based
- Voice
- Hands
- Body

External to body
- Paper & pencil
- Blackboard
- Computer screen
- Other electronic devices
- Paint, clay, stone, etc.
- Math manipulatives, blocks, etc.
- Musical instruments

Expressive Products

Bodily based
- Speech, song, chant
- Sign language, gestures
- Dance, marching, posing, etc.

External to body
- Inscriptions:
  - Written words, books, musical scores, etc.
  - Written math symbols, graphs, visuals
  - Text messages, web pages, computer games
- Other products:
  - Paintings, pottery, sculpture, buildings, etc.
  - A configuration of cubes, rods, blocks, etc.
  - Instrumental music

example, in American Sign Language. This form of language uses a different medium from oral speech, specifically, the hands, head, face, and body. It also utilizes a different sensory channel: sight. Thus, sign is a different expressive product from oral speech, employing different modalities. Yet at a higher level of abstraction, both products (oral speech and signing) are seen as language, a culturally based semiotic mode, developed and shared within specific communities.

In the realm of mathematics, an analysis utilizing the categories in Table 1.1 would suggest, for example, that "doing geometry" is a different experience, conceptually, for a learner who is working with pencil and paper versus a dynamic geometry tool instantiated on a computer. Indeed, a robust line of research has investigated the nature of students' experiences with computer-based dynamic geometry systems, examining how such systems afford different experiences from those afforded by paper and pencil (e.g., Lehrer & Chazan, 1998; Laborde, 1995; Mariotti, 2000;

Moreno-Armella, Hegedus, & Kaput, 2008). Further examples of the impact of modality are found in this volume. For example, the experiences of blind learners exploring the geometric concepts of area and volume would, of necessity, differ from those of learners with sight (Healy & Fernandes, Chapter 6, this volume), and people interacting verbally while imagining a three-dimensional geometric situation would enact still other modalities of expression (Moore-Russo & Viglietti, Chapter 7, this volume).

We have created this framework for multimodality not solely for the sake of systematicity but also to bring attention to elements of the process of making meaning that may be overlooked during research on mathematical practice. The field of education has moved beyond a simplistic model of teaching as the transmission of information, but it is still in the process of elaborating the complex means through which knowledge is constructed. We would argue that this process involves not just "interior" cogitation and "external" representations but a nuanced interaction of the body with shared social and cultural resources.

AFFORDANCES OF MODALITIES

This brings us to the question of what these different modalities make possible; that is, what are their affordances (Gibson, 1979)? Norman (1999), following Gibson, defines affordances as "the actionable properties between the world and an actor" (p. 39). A more detailed definition is offered by Rizzo (2006):

> Affordances are opportunities for actions available in the environments for individuals with proper sensory-motor abilities. They do not belong to the environment neither to the individual, but to their relationships. Affordances are emergent phenomena between distribution of energy in the environment and potential agents' behavior. (p. 239)

For example, a horizontal surface affords sitting (or eating or writing depending on the height and surface smoothness), and a teapot with a handle affords easy pouring of liquids, assuming an agent capable of the given motor actions. A computer-based dynamic geometry package offers different affordances to the learner than do graph paper and pencil, among which is the capability to "drag" and continuously transform an image of a geometric object (e.g., Laborde, 1995).

It is reasonable to assume that different modalities have developed at least in part because they have different affordances; in other words, the characteristics of each modality bring into play different possibilities for action and communication. In this section, we will examine the affordances, constraints, and complementarities among different modalities and expressive products, particularly those involved in doing mathematics.

Gesture and Speech

A particularly clear case of the complementary affordances of modalities can be found in examining oral speech and gesture. Kendon (1997), citing the *Oxford English Dictionary*, says gesture "refers to 'a movement of the body or of any part of it' that is 'expressive of thought or feeling'" (p. 109). McNeill (1992) was one of the first researchers to point out the complementary nature of speech and gesture, proposing that,

> Speech and gesture are elements of a single integrated process of utterance formation in which there is a synthesis of opposite modes of thought—global-synthetic and instantaneous imagery with linearly-segmented temporally extended verbalization. Utterances and thoughts realized in them are both imagery and language. (p. 35)

He also points out the way in which "each modality performs its own functions, the two modalities mutually supporting one another" (McNeill, 1992, p. 6). This is in contrast to a view of gesture as epiphenomenal, a simple illustration of what is being expressed through speech. A growing body of research has shown that gesturing has a much more significant role in reasoning, problem solving, and cognitive development than merely reinforcing speech (e.g., Alibali, Kita, & Young, 2000; Alibali, Spencer, Knox, & Kita, 2011; Arzarello, Paola, Robutti, & Sabena, 2009; Edwards, 2009; Gerofsky, 2010; Goldin-Meadow, 2003; Robutti, 2006; Roth, 2001).

Goodwin (2003) also notes the complementarity of speech and gesture, stating that speech is not simply a more "evolved" form of communication than gesture:

> [T]he way in which the structure of gesture differs markedly from language might reflect not the development of a new, more complex system from a simpler one, but instead a process of progressive differentiation within a larger set of interacting systems in which gesture is organized precisely to provide participants with resources that complement, and thus differ significantly from those afforded by language. (p. 23)

The potential for speech and gesture to convey different meanings has been examined experimentally by Goldin-Meadow and her colleagues (e.g., Goldin-Meadow, 2003, 2006; Goldin-Meadow, Kim, & Singer, 1999; Özçaliskan & Goldin-Meadow, 2009). Following McNeill (1992), Goldin-Meadow (2006) describes the different affordances of speech and gesture:

> Speech conveys meaning discretely, relying on codified words and grammatical devices. Gesture that accompanies speech conveys meaning holistically, relying on visual and mimetic imagery. Because gesture and speech employ such different forms of representation, the two modalities rarely contribute identical information to a message. (p. 36)

When different information is conveyed simultaneously by speech and gesture, Goldin-Meadow (2003) refers to this as a gesture-speech "mismatch." Research has shown that such mismatches (or nonredundancies) can signal a readiness to learn or an imminent cognitive transition. In these cases, a learner's gestures can express a change in understanding that has not yet been expressed in his or her speech; this phenomenon has been demonstrated in arithmetic, language learning, science, and the development of Piagetian conservations (Alibali, Church, Kita, & Hostetter, Chapter 2, this volume; Goldin-Meadow, 2003, 2006; Goldin-Meadow, Levine, & Jacobs, Chapter 3, this volume; Özçaliskan & Goldin-Meadow, 2009; Roth, 2001).

Affordances of Modalities for Mathematics

The different affordances of gesture and speech in teaching, learning, and doing mathematics have been investigated in this volume and elsewhere (e.g., Alibali, Spencer, Knox, & Kita, 2011; Arzarello, Paola, Robutti, & Sabena, 2009; Edwards, 2008, 2009; Ferrara, 2006; Gerofsky, 2010; Goldin-Meadow, Kim, & Singer, 1999; Nemirovsky, 2003; Nemirovsky & Ferrara, 2009; Núñez, 2009; Radford, 2009; Robutti, 2006; Roth, 2001; Valenzeno, Alibali, & Klatzky, 2003). We now turn to a range of additional modalities and expressive products that are commonly used in mathematics, with the goal of highlighting their important affordances and constraints. The current analysis considers the following characteristics drawn from prior research on gesture and speech (Goldin-Meadow, 2006; McNeill, 1992):

- Permanence: Does the modality result in an ephemeral or a more permanent expressive product or representation?
- Temporality: Is the modality or expression linear (where the message emerges sequentially in time), as in speech? Or is the expressive product perceived as a global whole (holistically or nonsequentially), as with gesture or inscribed imagery?
- Structure: Is the expression analytic, that is, made up of meaningful sub-units? Or is it a synthetic, non-decomposable whole?

Table 1.2 presents a set of modalities and expressive products that are commonly utilized in doing, teaching, and learning mathematics, as well as a brief summary of the affordances and other important characteristics of each.

TABLE 1.2 Affordances of Modalities/Expressive Products for Mathematics

Modality or Mode	Expressive Product	Characteristics/Affordances
Language	• Speech	Ephemeral, linear, analytic (composed of meaningful subunits). Prosody, rhythm and volume can give emphasis.
	• Written text	Permanent, linear, analytic (composed of meaningful subunits).
Formal Notations	• Written mathematical symbols	Permanent, generally linear, generally synthetic (although some symbols have meaningful subunits). Compressions of more complex/abstract ideas utilizing metonymy.
Visual Imagery	• Static graphs e.g., using Cartesian coordinates: an important conventional blend	Permanent. Global/holistic. Analytic—by convention, the parts are meaningful.
	• Static geometric diagrams	Permanent. Global (or holistic). Analytic. Iconic to elements of physical world, but intended to "point to" ideal forms.
	• Static conventional mathematical diagrams (other than graphs and geometric diagrams; e.g., Venn diagrams)	Permanent. Generally global/holistic. Have some characteristics of drawings and some of symbols. Nonarbitrary. Can be synthetic or analytic.
	• Marks drawn to highlight, emphasize or direct attention	Often spontaneous, can be permanent or ephemeral. Global/holistic. Synthetic.
Interactive Visual Imagery and/or Formal Notations	• Computer/calculator-based mathematics systems • Dynamic geometry systems, function graphers, etc.	Same characteristics as the components (mathematical symbols, graphs, etc.). However, the system affords instantaneous feedback and iterative exploration. Interaction via mouse & keyboard, or finger & touchscreen.
Motor Actions	• Gestures with empty hands	Ephemeral, global, synthetic.
	• Gestures holding artifact (pen, pointer etc.)	Affords more precise boundaries and point locations when gesturing.
	• Gesture involving an object in environment (table surface or edge, etc.)	The affordances of the object can be incorporated into the meaning of the gesture.
	• Other bodily actions/postures/gaze	Ephemeral, global, synthetic. Each with own particular affordances.

DISCUSSION

As we have seen, the term "multimodality" is polysemous: It has been used to refer to the range of sensory channels in the human organism, to the linkage of action and perception at the neural level, and to the use of multiple means of expression or representation, from formal systems such as speech to spontaneous and idiosyncratic bodily gestures. Given the rich collection of modalities and expressive products related to mathematics, we hope that an analysis such as the one presented here will encourage researchers and teachers alike to attend to the various affordances of different modalities. In fact, there is a small but growing body of research that supports the efficacy of appropriate gesturing, by teachers and learners, in mathematics instruction (e.g., Gerofsky, 2010; Goldin-Meadow, Kim, & Singer, 1999; Valenzeno, Alibali, & Klatzky, 2003). Similarly, designers of computer-based learning environments for mathematics have long discussed the power of multiple representations (graphs, symbols, tables, words, etc.). The perspective presented here on complementary affordances has the potential to both explain the synergy among representations and inspire designs that incorporate modalities that might have been overlooked.

Yet a closer look at embodiment and multimodality in mathematics perhaps raises more questions than it answers, at the level of both beginning and advanced mathematics. A common approach to mathematics teaching at the elementary school level includes the use of concrete physical materials or manipulatives. These materials, by design, offer certain affordances and constraints to the learner, and it is assumed that using them is an important component in the construction of new mathematical concepts. But how does this happen? How do physical actions with blocks or tiles provide grounding for understanding the mathematical concept and the conventional language and symbols associated with it? Does this embodied interaction with concrete materials persist as part of the student's mathematical knowledge? There are suggestions that such actions are re-externalized and expressed later through gesture (see e.g., Edwards, 2009); what other functions do gestures perform? What about more advanced mathematical ideas, which are not obviously grounded in physical action? How are the various modalities utilized in learning and doing mathematics at the secondary and university levels? Some of these questions are addressed in the current volume, but a full understanding of embodiment and multimodality in mathematics will require further sustained inquiry.

Embodied experience may be an essential component of knowledge, but it does not always support mathematical understanding (see e.g., Nuñéz, Edwards, & Matos, 1999). Additional research is needed to determine how embodiment may constrain or limit understanding of formal mathematics. After all, the requirements of the discipline for consistency and universality have resulted in structures that do not correspond neatly to everyday experience. Yet we would argue that the entire range of modalities, including those that are bodily based, are essential to learning, teaching, and practicing mathematics. Exploring embodiment and multimodality in diverse mathematical contexts, we believe, can only strengthen our understanding of mathematics education.

NOTE

1. This, of course, echoes McCullough's (1961) question in his article, "What Is a Number, That a Man May Know It, and a Man, That He May Know a Number?" without, in our case, objectifying number.

REFERENCES

Alibali, M. W., Kita, S., & Young, A. (2000). Gesture and the process of speech production: We think, therefore we gesture. *Language and Cognitive Processes, 15*(6), 593–613.

Alibali, M. W., Spencer, R. C., Knox, L., & Kita, S. (2011). Spontaneous gestures influence strategy choices in problem solving. *Psychological Science, 22*(9), 1138–1144.

Arcavi, A. (1994). Symbol sense: Informal sense-making in formal mathematics. *For the Learning of Mathematics, 14*(3), 24–35.

Arzarello, F., Paola, D., Robutti, O., & Sabena, C. (2009). Gestures as semiotic resources in the mathematics classroom. *Educational Studies in Mathematics, 70,* 97–109.

Arzarello, F., & Robutti, O. (2008). Framing the embodied mind approach within a multimodal paradigm. In L. English (Ed.), *Handbook of international research in mathematics education* (pp. 720–749). New York, NY: Taylor & Francis.

Arzarello, F., & Robutti, O. (2010). Multimodality in multi-representational environments. *ZDM—The International Journal of Mathematics Education, 42*(7), 715–731.

Barsalou, L. (2008). Cognitive and neural contributions to understanding the conceptual system. *Current Directions in Psychological Science, 17*(2), 91–95.

Bateman, J. (2009). Discourse across semiotic modes. In J. Renkema (Ed.), *Discourse, of course: An overview of research in discourse studies* (pp. 55–66). Amsterdam, the Netherlands: John Benjamins Publishing.

Bateson, G. (2000). *Steps to an ecology of mind: Collected essays in anthropology, psychiatry, evolution, and epistemology.* Chicago, IL: University of Chicago Press.

Cajori, F. (1993). *A history of mathematical notations* (Vol. 1). Mineola, NY: Dover Publications.

Clark, A. (2001). *Being there: Putting brain, body, and world together again.* Cambridge, MA: Bradford Press.

Dubinsky, E., & Harel, G. (1992). The nature of the process conception of function. In G. Harel & E. Dubinsky (Eds.), *The concept of function: Aspects of epistemology and pedagogy* (Vol. 25, pp. 85–106). Washington, DC: Mathematical Association of America (MAA Notes).

Edwards, L. D. (1995). Microworlds as representations. In A. A. diSessa, C. Hoyles, R. Noss, & L. D. Edwards (Eds.), *Computers and exploratory learning* (pp. 127–154). New York, NY: Springer.

Edwards, L. D. (2008). Conceptual integration, gesture and mathematics. In O. Figueras & A. Sepúlveda (Eds.), *Proceedings of the joint meeting of the 32nd conference of the International Group for the Psychology of Mathematics Education, and the XX North American Chapter* (Vol. 2, pp. 423–430). Morelia, Mexico: University of Michoacan Press.

Edwards, L. D. (2009). Gestures and conceptual integration in mathematical talk. *Educational Studies in Mathematics, 70*(2), 127–141.

Fauconnier, G., & Turner, M. (2002). *The way we think: Conceptual blending and the mind's hidden complexities.* New York, NY: Basic Books.

Ferrara, F. (2006). Remembering and imagining: Moving back and forth between motion and its representation. In J. Novotná, H. Moraová, M. Krátká, & N. Stehlíková (Eds.), *Proceedings of the 30th conference of the International Group for the Psychology of Mathematics Education* (Vol. 3, pp. 65–72). Prague, Czech Republic: Charles University Press, Faculty of Education.

Font, V., Godino, J., Planas, N., & Acevedo, J. (2009). The existence of mathematical objects in the classroom discourse. In V. Durand-Guerrier, S. Soury-Lavergne, & F. Arzarello (Eds.), *Proceedings of the 6th Congress of the European Society for Research in Mathematics Education* (pp. 984–995). Lyon, France: INRP. Available at http://www.inrp.fr/editions/editions-electroniques/cerme6/.

Font, V., Godino, J., Planas, N., & Acevedo, J. (2010). The object metaphor and synecdoche in mathematics classroom discourse. *For the Learning of Mathematics, 30*(1), 15–19.

Gallese, V., & Lakoff, G. (2005). The brain's concepts: The role of the sensory-motor system in conceptual knowledge. *Cognitive Neuroscience, 22,* 455–479.

Gerofsky, S. (2010). Mathematical learning and gesture: Character viewpoint and observer viewpoint in students' gestured graphs of functions. *Gesture, 10*(2–3), 321–343.

Gibson, J. J. (1979). *The ecological approach to visual perception.* Boston, MA: Houghton Mifflin.

Goldin-Meadow, S. (2003). *Hearing gesture: How our hands help us think.* Cambridge, MA: Harvard University Press.

Goldin-Meadow, S. (2006). Talking and thinking with our hands. *Current Directions in Psychological Science, 35*(1), 35–39.

Goldin-Meadow, S., Kim, S., & Singer, M. (1999). What the teacher's hands tell the student's mind about math. *Journal of Educational Psychology, 91*(4), 720–730.

Goodwin, C. (2003). The body in action. In J. Coupland & R. Gwyn (Eds.), *Discourse, the body, and identity* (pp. 19–42). New York, NY: Palgrave Macmillan.

Hutchins, E. (2005). Material anchors for conceptual blends. *Journal of Pragmatics, 37,* 1555–1577.

Johnson, M. (1987). *The body in the mind.* Chicago, IL: University of Chicago Press.

Johnson, M. (2007). *The meaning of the body: Aesthetics of human understanding.* Chicago, IL: University of Chicago Press.

Kendon, A. (1997). Gesture. *Annual Review of Anthropology, 26,* 109–128.

Kieran, C. (1992). The learning and teaching of school algebra. In D. Grouws (Ed.), *Handbook of research on mathematics teaching and learning* (pp. 390–419). New York, NY: Macmillan.

Kilpatrick, J. (1992). A history of research in mathematics education. In D. Grouws (Ed.), *Handbook of research on mathematics teaching and learning* (pp. 3–38). New York, NY: Macmillan.

Kling, J. W., & Riggs, L. A. (1971). *Experimental psychology.* New York, NY: Holt, Rinehart and Winston.

Kress, G. (2001a). *Multimodal discourse: The modes and media of contemporary communication.* London, England: Arnold Publishers.

Kress, G. (2001b). Sociolinguistics and social semiotics. In P. Copely (Ed.), *The Routledge companion to semiotics and linguistics* (pp. 66–82). London, England: Routledge.

Kress, G. (2005). Gains and losses: New forms of texts, knowledge, and learning. *Computers and Composition, 22,* 5–22.

Kress, G., & Van Leeuwen, T. (2002). Colour as a semiotic mode: Notes for a grammar of colour. *Visual Communication, 1*(3), 343–368.

Laborde, C. (1995). Designing tasks for learning geometry in a computer-based environment: The case of Cabri-geometre. In L. Burton & B. Jaworski (Eds.), *Technology in mathematics teaching: A bridge between teaching and learning* (pp. 35–67). Bromley, England: Chartwell-Bratt.

Lakoff, G. (1993). The contemporary theory of metaphor. In A. Ortony (Ed.), *Metaphor and thought* (2nd ed., pp. 203–204). Cambridge, England: Cambridge University Press.

Lakoff, G., & Johnson, M. (1999). *Philosophy in the flesh.* New York, NY: Basic Books.

Lakoff, G., & Núñez, R. (2000). *Where mathematics comes from: How the embodied mind brings mathematics into being.* New York, NY: Basic Books.

Lehrer, R., & Chazan, D. (Eds.). (1998). *Designing learning environments for developing understanding of geometry and space.* New York, NY: Routledge.

Lerman, S. (2000). The social turn in mathematics education research. In J. Boaler (Ed.), *Multiple perspectives on mathematics teaching and learning* (pp. 19–44). Westport, CT: Ablex.

Mariotti, M. A. (2000). Introduction to proof: The mediation of a dynamic software environment. *Educational Studies in Mathematics, 44*(1/2), 25–53.

Matz, M. (1982). Towards a process model for high school algebra errors. In D. Sleeman & J. S. Brown (Eds.), *Intelligent tutoring systems* (pp. 25–50). New York, NY: Academic Press.

McCullough, W. (1961). What is a number, that a man may know it, and a man, that he may know a number? *General Semantics Bulletin,* 26–27, 7–18.

McNeill, D. (1992). *Hand and mind: What gestures reveal about thought.* Chicago, IL: University of Chicago Press.

McNeill, D. (Ed.). (2005). *Gesture and thought.* Chicago, IL: University of Chicago Press.

Moreno-Armella, L., Hegedus, S., & Kaput, J. (2008). From static to dynamic mathematics: Historical and representational perspectives. *Educational Studies in Mathematics, 68*, 99–111.

Nemirovsky, R. (2003). Three conjectures concerning the relationship between body activity and understanding mathematics. In N. A. Pateman, B. J. Dougherty, & J. T. Zilliox (Eds.), *Proceedings of 27th conference of the International Group for the Psychology of Mathematics Education* (Vol. 1, pp. 105–109). Honolulu, HI: University of Hawaii Press.

Nemirovsky, R., & Ferrara, F. (2009). Mathematical imagination and embodied cognition. *Educational Studies in Mathematics, 70*(2), 159–174.

Norman, D. A. (1999). Affordance, conventions, and design. *interactions, 6*(3), 38–43.

Norris, S. (2004). *Analyzing multimodal interaction: A methodological framework.* New York, NY: Routledge.

Núñez, R. (2008). Mathematics, the ultimate challenge to embodiment: Truth and the grounding of axiomatic systems. In P. Calvo & T. Gomila (Eds.), *Handbook of cognitive science: An embodied approach* (pp. 333–353). Amsterdam, the Netherlands: Elsevier.

Núñez, R. (2009). Gesture, abstraction, and the embodied nature of mathematics. In W.-M. Roth (Ed.), *Mathematical representation at the interface of body and culture* (pp. 309–328). Charlotte, NC: Information Age Publishing.

Núñez, R., Edwards, L. D., & Matos, J. (1999). Embodied cognition as grounding for situatedness and context in mathematics education. *Educational Studies in Mathematics, 39*(1–3), 45–65.

Özçaliskan, S., & Goldin-Meadow, S. (2009). When gesture-speech combinations do and do not index linguistic change. *Language & Cognitive Processes, 24*(2), 190–217.

Quine, W. V. (1950). Identity, ostension, and hypostatis. *The Journal of Philosophy, 47*(22), 621–633.

Radford, L. (2009). Why do gestures matter? Sensuous cognition and the palpability of mathematical meanings. *Educational Studies in Mathematics, 70*(3), 111–126.

Radford, L. (2011). Embodiment, perception and symbols in the development of early algebraic thinking. In B. Ubuz (Ed.), *Proceedings of the 35th conference of the International Group for the Psychology of Mathematics Education* (Vol. 4, pp. 17–24). Ankara, Turkey: PME.

Reddy, M. J. (1979). The conduit metaphor: A case of frame conflict in our language about language. In A. Ortony (Ed.), *Metaphor and thought* (pp. 284–310). Cambridge, England: Cambridge University Press.

Reid, C. (1996). *Hilbert.* Berlin, Germany: Springer.

Rizzo, A. (2006). The origin and design of intentional affordances. In J. Carroll, S. Bødker, & J. Coughlin (Eds.), *Proceedings of the 6th conference on Designing Interactive Systems* (pp. 239–240). New York, NY: Association of Computing Machinery.

Robutti, O. (2006). Motion, technology, gesture in interpreting graphs. *The International Journal for Technology in Mathematics Education, 13*(3), 117–126.

Roth, W.-M. (2001). Gestures: Their role in teaching and learning. *Review of Educational Research, 71*, 365–392.

Sfard, A. (1994). Reification as the birth of metaphor. *For the Learning of Mathematics, 14*(1), 44–55.

Struik, D. J. (1986). The sociology of mathematics revisited: A personal note. *Science and Society, 50*, 280–299.

Tall, D. O., Thomas, M., Davis, G., Gray, E. M., & Simpson, A. (2000). What is the object of the encapsulation of a process? *Journal of Mathematical Behavior, 18*, 223–241.

Valenzeno, L., Alibali, M. W., & Klatzky, R. (2003). Teachers' gestures facilitate students' learning: A lesson in symmetry. *Contemporary Educational Psychology, 28*(2), 187–204.

SECTION I

GESTURE AND EMBODIMENT IN EARLY MATHEMATICS

CHAPTER 2

EMBODIED KNOWLEDGE IN THE DEVELOPMENT OF CONSERVATION OF QUANTITY

Evidence from Gesture

Martha W. Alibali
University of Wisconsin, Madison

R. Breckinridge Church
Northeastern Illinois University

Sotaro Kita
The University of Warwick

Autumn B. Hostetter
Kalamazoo College

Understanding quantity is one of the foundations of mathematical thinking. Piaget (1941/1952) recognized this in his seminal work, *The Child's Conception of Number*, in which he addresses conservation of quantity in the first

Emerging Perspectives on Gesture and Embodiment in Mathematics, pages 27–49

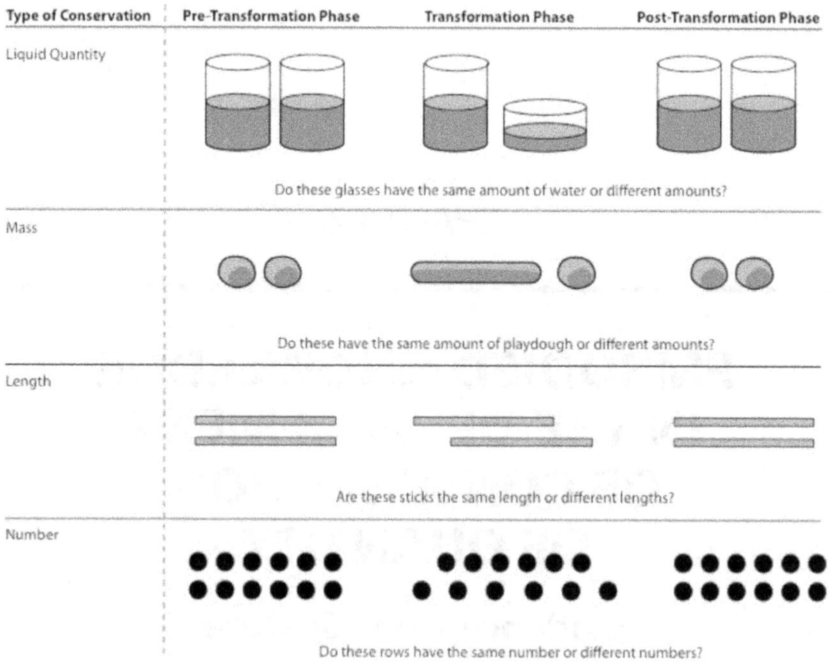

Figure 2.1 Schematic of conservation task.

two chapters. Conservation of quantity is the understanding that quantities remain invariant under certain sorts of transformations that alter perceptual appearance (see Figure 2.1). Consider, for example, transformations of discrete (countable) quantities, such as sets of checkers or blocks. In a classic Piagetian conservation task, the experimenter might alter the physical position of a row of checkers by spreading them out so that the row is longer and the spaces between checkers are larger, making the row appear more numerous. Similarly, consider transformations of continuous (noncountable) quantities, such as liquid or sand. An experimenter might pour a liquid into a container that is shorter and wider, making the amount of liquid appear less.

A full understanding of conservation requires understanding that, although a transformation alters appearance, it does not in fact alter quantity. A child who conserves understands that, despite the misleading appearance of the transformed object, the quantity has not changed. In contrast, children who do not yet conserve are swayed by the perceptual appearance of a transformed quantity. Conserving children's responses reveal that they use a variety of operations basic to mathematical understanding (including one-to-one correspondence, counting, reversibility, identity, and quantitative comparison) to justify why quantities remain the same after transformation (e.g., Rains, Kelly, & Durham, 2008). Understanding that quantity is conserved under certain sorts of transformations can be characterized in terms of progress from a perceptually based approach to an abstract, logic-based, or "cognitive" approach to understanding quantity (e.g., Piaget, 1941/1952; Schultz, 1998).

A great deal of research on children's understanding of conservation has focused on patterns of change over development. Piaget viewed children's performance on conservation tasks as indicative of their stage of cognitive development and consequently focused on discontinuities or sudden shifts in children's performance (Piaget, 1941/1952). Other investigators have challenged Piaget's views, criticizing on numerous grounds his notion of "stage" and his claim that stage-related cognitive processes constrain children's learning. Studies have revealed that many children conserve at younger ages than predicted by Piaget's theory (e.g., McGarrigle & Donaldson, 1974), that children can be trained to conserve (e.g., Gelman, 1969), that social and pragmatic factors affect children's conservation performance (e.g., Siegal & Waters, 1988), that there are cross-cultural variations in conservation performance (e.g., Irvine, 1978), and that, on close inspection, transitions in conservation understanding are not as sudden or abrupt as Piaget believed (Church & Goldin-Meadow, 1986).

Starting in the 1970s, research on conservation took an information-processing approach, focusing on the fine structure of children's performance, such as their encoding of the conservation tasks and their patterns of strategy use (e.g., Klahr & Wallace, 1973; Siegler & Robinson, 1982). This work informed the development of computer simulations of children's conservation performance and learning, including production system models (e.g., Klahr & Wallace, 1976) and connectionist models (Schultz, 1998), which sought to explain both gradual and discontinuous changes in children's performance. Other contemporary research has focused on accounting for patterns of change from the perspectives of catastrophe theory (van der Maas & Molenaar, 1992) and dynamic systems theory (van Geert, 1998). Still other recent research has focused on understanding the brain bases of performance on conservation tasks (Houdé et al., 2011; Zhang et al., 2008).

One fundamental question addressed across theoretical frameworks is how children acquire an understanding of conservation. One likely possibility is that children observe and implement transformations of objects and substances in their everyday lives and play; as they do so, they begin to reason about the effects of such transformations. For example, children may notice that pouring liquid into a wider container also alters the height of the liquid (compensation), or they may notice that a transformation such as spreading out a row of coins can be "undone" by squishing them back together (reversibility). Indeed, evidence shows that experience with quantities, and in particular experience manipulating quantities, can promote understanding of conservation. For example, Mexican children who grew up in pottery-making families showed earlier understanding of conservation of substance (tested with clay) than control children who did not grow up in pottery-making families (Price-Williams, Gordon, & Ramirez, 1969).

Manipulating concrete quantities is part of children's natural, everyday experience, and many researchers have suggested that such experience provides a foundation for later understanding of mathematical principles (e.g., Klahr & Wallace, 1976; Schliemann & Carraher, 2002). If this is the case, it suggests that children's initial knowledge of conservation might be embodied, in the sense that it is based on perception and action in the physical world (see e.g., Barsalou, 2008; Glenberg, 2010; Wilson, 2002). Over time, such knowledge might become more abstract or linked to more abstract knowledge structures, such as rules. It is worth noting that this progression from more perceptually based approaches to more abstract, logically based, "cognitive" approaches over developmental time is well aligned with Piaget's theoretical claims about the sensorimotor origins of knowledge and the gradual abstraction of knowledge over developmental time. Moreover, this progression has been documented in numerous studies of conservation development (e.g., Piaget, 1941/1952; Schultz, 1998). In this chapter, we review recent research on children's understanding of conservation, with an eye toward understanding the embodied nature of such knowledge and the implications of an embodied perspective for understanding processes of knowledge change.

EMBODIED KNOWLEDGE OF CONSERVATION

A fundamental claim of theories of embodied cognition is that human cognitive processes are rooted in perceptual and physical interactions of the human body with the physical world (e.g., Barsalou, 2008; Glenberg, 2010; Wilson, 2002). A related claim is that the human conceptual system is based on fundamental aspects of human experience (such as common actions, spatial relations, and bodily experiences), which provide an embodied foundation for abstract concepts via metaphoric structuring (e.g., Lakoff & Johnson, 1980; Lakoff & Núñez, 2001).

Many investigators (including many of those in this volume) have argued specifically that mathematical thinking is embodied (e.g., Lakoff & Núñez, 2001). Embodied perspectives on mathematics and mathematics instruction have emphasized the body-based conceptual metaphors that underlie mathematical thinking (e.g., Núñez, Edwards, & Matos, 1999) and the importance of perceptual processes and actions (both real and simulated) in mathematical thinking (e.g., Alibali & Nathan, 2012; Ben-Zeev & Star, 2001; Landy & Goldstone, 2007).

What empirical evidence suggests that reasoning about conservation is embodied? One possible source of evidence can be found in children's gestures. Indeed, many investigators have suggested that gestures can be taken as evidence that the knowledge expressed in those gestures is embodied (Gibbs, 2006; McNeill, 2005; Núñez, 2005). We suggest that embodied reasoning about conservation might be manifested in the gestures that children produce when they explain their reasoning about the conservation tasks. The basic idea is that spontaneous gestures reflect the perceptual and physical bases of the knowledge they express.

Gestures Reflect Embodied Knowledge

Hostetter and Alibali (2008) have articulated a theory about how gesture might reflect embodied knowledge, which they have called the Gesture as Simulated Action (GSA) framework. Briefly, the GSA framework holds that speakers produce gestures when they simulate actions and perceptual states in service of speaking. A simulation can be defined as a re-creation of the neural states involved in acting or perceiving without actually acting or perceiving (see also Gallese & Lakoff, 2005). During simulation, motor and premotor areas of the brain are activated as they are when actually acting or perceiving. For example, imagining oneself performing an action activates the motor cortex, the cerebellum, and the basal ganglia (Jeannerod, 2001). According to the GSA framework, when this motor activation exceeds a threshold (the height of which is determined by individual and social factors), it is realized in overt movements, which we recognize as gestures.

From this perspective, gestures in children's explanations of conservation tasks would suggest that children utilize perception- and action-based thinking in reasoning about conservation. In the following sections, we review research on gesture in children's conservation explanations as a source of evidence about the embodied nature of children's understanding of conservation and its development. We first consider the nature of the gestures that children produce when they explain conservation tasks: how frequent they are, what kinds of information they express, and how that information aligns with the information that children express in speech.

Children Gesture at High Rates When Solving Conservation Problems

Do children commonly produce gestures when they solve and/or speak about conservation problems? Indeed, they do. In a study conducted by Church and Goldin-Meadow (1986, Study 1), children between the ages of 5 and 8 years participated in a series of six conservation tasks measuring their understanding of conservation of continuous quantities (liquid quantity and length) and discontinuous quantities (number). Each task contained three phases: a pretransformation phase, a transformation phase, and a posttransformation phase. In the pretransformation phase, the child was presented with two sets with equal quantity (i.e., two identical glasses of water, two aligned sticks of equal length, or two aligned rows with the same number of checkers) and asked to compare their quantities (e.g., they were asked, "Do these two [glasses] [sticks] [rows] have the same [amount of water] [length] [number]?") In the transformation phase, one of the task objects (i.e., one glass of water, one stick, or one row of checkers) was transformed (e.g., the glass of water was poured into a short round dish, the stick was moved horizontally to the side, or the row of checkers was spread out). All of the transformations were performed directly in front of the child. The child was then again asked to compare the two quantities (e.g., "How about now? Do the glass and the dish have the same amount of water?"). Once the child provided the conservation judgment (either "Yes, they have the same amount," indicating conservation of quantity, or "No, they have different amounts," indicating lack of conservation of quantity), he or she was asked to provide an explanation for why the objects were the same or different in amount. In the posttransformation phase, the transformed quantity was returned to its original state (the water poured back into its original container, the stick pushed back to be aligned with the other stick, or the row of checkers compressed to be the same length and aligned with the other row of checkers), and the child was again asked to compare the quantities. This entire process was videotaped.

The critical data that were analyzed in this study were drawn from children's judgments and explanations. As children explained their judgments, they often produced gestures that provided information similar to that contained in their verbal explanations. For example, one child claimed that the water in the dish was "shorter" than the water in the glass (justifying why she judged the comparison objects to be different in quantity) and, while making this claim, demarcated with her hand the height of the dish and the height of the glass (highlighting in gesture the difference in height of the two containers).

Church and Goldin-Meadow (1986) reported that all but 1 of 28 participants (i.e., 96.4%) gestured on at least one of the six conservation tasks. The children varied in whether they understood that the transformations did not alter quantity, with some children (conservers) correctly judging that none of the six transformations altered quantity, some children (non-conservers) incorrectly judging that all of the transformations did so, and other children (partial conservers) showing variable performance. However, regardless of performance level, the majority of children's explanations included gestures: non-conservers produced gestures in an average of 5.2

of 6 explanations (i.e., 87%), partial conservers produced gestures in an average of 5 of 6 explanations (83%), and conservers produced gestures in an average 4.8 of 6 explanations (i.e., 80%). Thus, children often produced gestures regardless of whether they solved the conservation problems correctly. In another study using the identical conservation task procedures, Church (1999) reported that, for a sample of non- and partial conservers, 95% of children's responses on a conservation pretest included gestures. Thus, if gestures do in fact manifest embodied knowledge, children's frequent production of gesture during conservation tasks suggests that their knowledge of conservation is indeed embodied.

Children's Gestures Convey Perceptual Features and Actions

Children commonly express their knowledge about conservation through their gestures. But what kinds of information do they express in those gestures? In a study by Mainela-Arnold, Evans, and Alibali (2006), the majority of children's gestured explanations of conservation tasks focused on perceptual features of the task objects, specifically, features that could be perceived at the moment of explanation. Children's gestured responses included explanations that focused on a single perceptual attribute of one of the task objects (e.g., on a task in which water was poured from a tall glass into a shorter dish, a focus on the short height of the dish), explanations that focused on the same perceptual attribute on both task objects (e.g., on the same task, the tall height of the glass and the short height of the dish), and explanations that focused on compensating perceptual attributes on one or both task objects (e.g., the height and width of the dish).

Note that in many cases, gestures that express perceptual attributes of conservation task objects have both iconic (i.e., depictive) and deictic (i.e., pointing) components. We suggest that deictic gestures reflect the grounding of children's explanations in the physical world, and iconic gestures reflect sensorimotor experience and, specifically, the simulation of actions or perceptual states (see Alibali & Nathan, 2012, for discussion of these ideas). For example, in a liquid conservation task, a child who describes the height of a glass by holding a flat palm facing down at the top edge of the glass displays in gesture an iconic representation of the glass's height. This representation is also deictic because it is produced on the glass, and in this way it indicates or "points" to the glass. As a second example, a child might describe the round shape of a flattened ball of clay by tracing its circular shape with her finger. This gesture is iconic in that it depicts the round shape of the mass of the clay, and it is deictic because it points to that specific mass of clay. In other cases, gestures that express perceptual attributes of the task objects are solely deictic. For example, a child might simply point to the water level in a glass to indicate that level or point to a space between checkers in describing the density of checkers in a row.

Some researchers have asked children to respond to conservation tasks after the objects have been removed from view. In one such study, children produced more solely deictic gestures and more gestures with both iconic and deictic components when the task objects were present than when they were absent (Ping & Goldin-Meadow, 2010). Thus, not surprisingly, the presence of the objects appears to "afford" deictic gestures. However, even when objects are absent, children continue to produce gestures when explaining their reasoning about conservation.

These examples illustrate speakers' use of gestures to connect their verbal utterances to their referents—or to imaginary or simulated referents—in the physical environment (Alibali & Nathan, 2012). Children's gestures in typical (objects-present) conservation tasks ground their conservation reasoning in the physical environment, as it exists at the moment of explanation, with physically present task objects such as glasses, water, clay, and so forth, in particular perceptual configurations. Such gestures may serve to highlight particular features of the immediate perceptual display, both for purposes of communicating to the experimenter and for children's own reasoning about the task (see Alibali & Kita, 2010). Children's gestures to imagined objects (e.g., in conservation tasks with objects removed from view) suggest that, in some cases, speakers simulate perceptual properties of simulated task objects (see Hostetter & Alibali, 2008).

Children's gestures also frequently depict or simulate actions, including the actions that the experimenter performed during the task (e.g., pouring the water from the glass to the dish, flattening the clay) and actions that the children or experimenter could perform on the task objects (e.g., reversing the transformation by pouring the water from the dish back into the original glass or actually altering quantity by pouring some water out of the dish). These gestures reflect aspects of the task that are not perceptually present at the moment of explanation (Mainela-Arnold et al., 2006). Simulating such actions via gesture may help speakers and listeners envision such actions and reason about their consequences.

Gesture as a Means of Exploring the Conservation Task

Why do conservation tasks so consistently evoke gestures? One possibility is that producing gestures is actually a means of reasoning about the tasks. Alibali, Kita, and Young (2000) argued that children's gestures are involved in the conceptual planning of their conservation explanations. They examined responses in which children focused on perceptual attributes of the task objects, both when children were asked to explain their conservation judgments and when children were simply asked to describe the task objects. Children produced more gestures that expressed perceptual features of the task objects (e.g., depicting the width of a container using a C-shaped handshake, demarcating the length of a row of blocks with flat palms at each end of the row) when they were explaining than when they were describing. Thus, explaining quantity judgments elicits gestures that highlight perceptual aspects of the conservation tasks.

Aligned with this view, we suggest that the action of producing gestures may actually influence the information that children encode and represent when solving conservation tasks. Children's behavior in task explanations suggests that they sometimes use gesture to "explore" aspects of the conservation tasks, and the information they "discover" using gestures may serve as candidate ideas to be incorporated into their verbal explanations. Some examples can illustrate how this might occur. At times, children highlight perceptual dimensions of the task objects in gestures just moments before they express those dimensions in speech. For example, in explaining his judgment about a task in which water had been poured from a tall, thin glass to a short, wide dish, one child in the dataset reported on by Alibali et al. (2000) said, "This is just, umm, really fat" and produced a gesture depicting the width of the dish with both hands in "C" shapes at its sides (see Figure 2.2). The child began his gesture at the outset of his utterance, on the word "this," well before he verbally expressed the idea of the width of the container ("fat"). It is possible that the boy's gesture may have actually

Figure 2.2 Response in which a child expresses information about the width of a container in gesture moments before he expresses that same information in speech (see text). This response is an example of a gesture-speech match.

focused his attention on the width of the dish and highlighted that information for expression in his explanation.

At other times, children express aspects of the conservation tasks (e.g., features of the task objects, real or possible actions on the task objects) in gesture that they never express in their spoken explanations. For example, in explaining her judgment about a task in which a ball of clay had been rolled into a long, thin "sausage," one girl (also from the Alibali et al. [2000] dataset) said, "Because that one's long." While saying the word "because," the child produced a gesture with both palms down, making a rolling motion in front of her body, simulating the rolling transformation that the experimenter had performed a few moments before (Figure 2.3A). Then while saying "that one's," the child produced a sweeping motion with her right palm along the clay "sausage," delineating its length (Figure 2.3B). In this example, the child expressed information about the transformation in gesture but never expressed that information in speech. In this case, the child seemed to "try out" in gesture the rolling action that she saw the experimenter perform, but she ultimately chose to say (and gesture) something different—namely, the length of the transformed object, which she also highlighted in gesture. Responses such as this one, in which gestures express information that is not expressed in the accompanying speech, were first studied by Church and Goldin-Meadow (1986) and termed *gesture-speech mismatches.*

In gesture-speech mismatch responses, gestures appear to explore aspects of the task and the task objects that could be useful for answering the experimenter's question, but not all of gesture's "discoveries" are reflected in children's verbal explanations. In support of this view, Goldin-Meadow, Alibali, and Church (1993) reported that children who frequently produced mismatches in an equation-solving task tended to express a greater number of different strategies (across speech and gesture) than children who rarely produced mismatches, and many of those strategies were expressed only in gestures and never in speech. In conservation, Alibali et al. (2000) found that children expressed ideas uniquely in gesture more often when they explained their conservation judgments than when they described the conservation task objects. This may be because the explanation task requires children to consider a wider range of ideas than the description task. Gestural exploration of aspects of the conservation tasks thus appears to be a way to introduce new ideas into the cognitive system.

Hence, children appear to use gestures to explore aspects of the task objects for use in reasoning about the conservation tasks. From this perspective, gesture is a form of action that may not only reflect aspects of children's knowledge but may also increase activation of certain types of information or even bring new information into the mix. If this is the case, it suggests that people should reason differently about tasks when gesture is allowed than

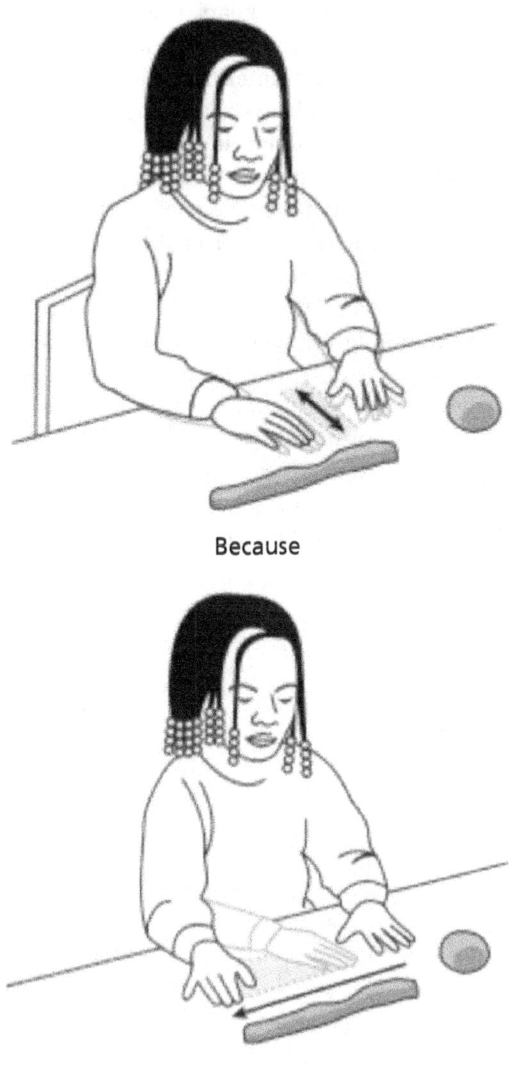

Figure 2.3 Response in which a child expresses information about the transformation (rolling) in gesture but not in speech (panel A, see text). This response is an example of a gesture-speech mismatch. The child also expresses information about the length of the sausage in both gesture and speech.

when it is prohibited. Indeed, a recent study of adults' reasoning about gear movements indicated that adults used a different mix of strategies to solve the problems when gesture was allowed and when gesture was prohibited (Alibali, Spencer, Knox, & Kita, 2011). When gesture was allowed, people tended to simulate the actions of the gears, whereas when gesture was prohibited, they were more likely to reason about the number of gears.

Along similar lines, recent experimental evidence suggests that producing gestures also influences children's reasoning about conservation. Alibali and Kita (2010) studied this issue using a gesture prohibition paradigm with 5- to 7-year-old children. In an initial set of three conservation tasks, all of the children were allowed to gesture as they explained their judgments. Then in a second set of three tasks, half of the children were prohibited from gesturing by placing their hands in a cloth muff. Children who were allowed to gesture on the second set of tasks were more likely to express information about the immediate perceptual state of the task objects and less likely to express information that was not perceptually present at the moment of explanation than were children who were prevented from gesturing (see Figure 2.4). So, producing gestures appears to highlight or lend salience to perceptually present information.

Thus, it appears that producing gestures affects the knowledge that children activate when they explain their reasoning about conservation tasks,

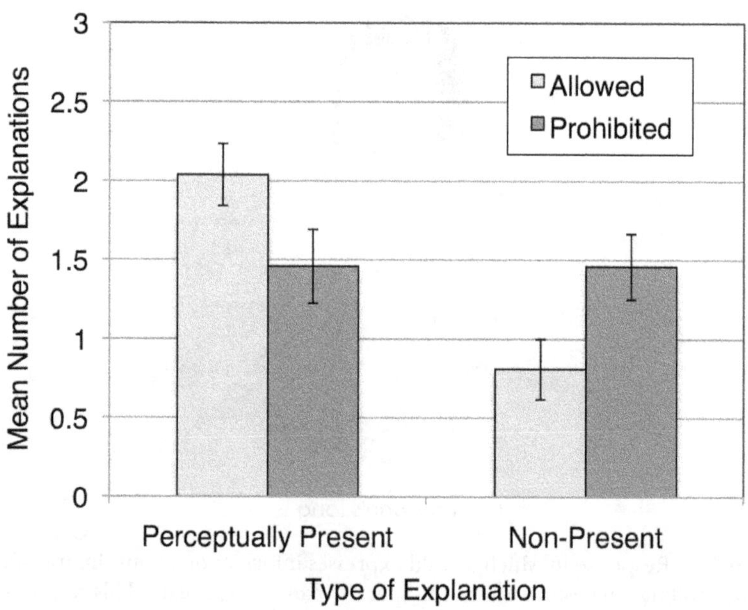

Figure 2.4 Mean number of perceptually present and nonpresent explanations produced in the second set of tasks by children for whom gesture was allowed and for children for whom gesture was prohibited [data from "Gesture highlights perceptually present information for speakers" by M.W. Alibali & S. Kita (2010). *Gesture, 10*(1), 3–28. With kind permission of John Benjamins Publishing Company, Amsterdam/Philadelphia. www.benjamins.com].

by both exploring aspects of the conservation task and highlighting perceptually present information. If gesture affects performance in these ways, it opens the possibility that gesture may also influence change in knowledge about conservation, at least in some cases. We turn next to research that shows a correlation between production of gestures that "mismatch" speech and change in children's knowledge of conservation.

EMBODIED KNOWLEDGE AND PROCESSES OF CHANGE IN CONSERVATION UNDERSTANDING

The studies reviewed thus far indicate that children typically gesture when they explain conservation tasks and appear to use gesture to explore perceptual features and possible actions in the conservation task. It is possible that children's gestures, used as they are to explore the tasks and express and communicate their knowledge, may actually be involved in the mechanism of knowledge change in conservation understanding. In this section, we consider two classes of mechanisms for how gesture may be involved in change in conservation: (a) internal mechanisms that focus on how children's gestures reflect and influence knowledge change, and (b) externally oriented mechanisms that involve others detecting and responding to the embodied understanding that children express in their gestures, thus contributing to the social construction of knowledge change.

Children's Gesture Production Reflects and May Influence Knowledge Change

Gesture-Speech Mismatches Index Transitional Knowledge About Conservation

Gesture-speech mismatches have been documented in learners' problem explanations, not only in conservation but in a wide range of other cognitive tasks, including language development (Özçaliskan & Goldin-Meadow, 2009), mathematical equations (Perry, Church, & Goldin-Meadow, 1988), balance tasks (Pine, Lufkin, & Messer, 2004), and science explanations (Roth, 2002). Across several tasks, there is evidence that children who frequently produce gesture-speech mismatches when explaining a particular concept are in a state of transitional knowledge about that concept (e.g., Perry et al., 1988; Pine et al., 2004). Church and Goldin-Meadow (1986), studying a sample of partial conservers, found that children who produced predominantly gesture-speech mismatches in their conservation explanations prior to instruction were more likely to profit from instruction about conservation (i.e., they were more likely to demonstrate conservation understanding after instruction) than were children who produced few mismatches. Thus, frequent mismatches between speech and gesture in children's task explanations index readiness to learn.

What is it about mismatching gesture that would indicate that a child is ready to learn about conservation? Gesture-speech mismatches reflect the simultaneous activation of multiple, alternative representations, including embodied knowledge representations and verbally coded ones, at the moment a child is explaining the task. According to Church (1999), "When the simultaneous activation of these multiple representations occurs, this marks the process of integration or reorganization" (p. 338). From this perspective, gesture-speech mismatches presage the integration of representations that is critical to move beyond perceptually based reasoning to a more abstract understanding of conservation of quantity.

Gesture-Speech Mismatches Are More Predictive of Knowledge Change Than Other Indices of Transition

Not only do gesture-speech mismatches index readiness to learn—they are also better at doing so than verbal explanations that reflect multiple approaches to conservation problems. Church (1999) compared three potential predictors of learning from a lesson about conservation: (a) number of gesture-speech mismatches produced at pretest, (b) number of different types of "arguments" (different rationales, such as transformation, comparison of a single dimension, compensation, etc.) expressed in verbal explanations across explanations on the conservation pretest, and (c) number of pretest verbal explanations that included multiple conservation "arguments" within a single verbal explanation (e.g., comparison and transformation arguments in the same verbal explanation). All three indices predicted learning, but when all three were included in the same statistical model, only gesture-speech mismatches were a significant predictor of children's success on a conservation posttest.

It appears that gesture-speech mismatches are a robust predictor of transition in conservation understanding because they reflect a knowledge state in which children explore and activate embodied representations alongside more abstract, verbally coded representations. As children experience and act on the world, their perception- and action-based understanding of quantity invariance may grow and change slightly in advance of their verbal understanding. This asynchrony may not only indicate a state of readiness to change but may in fact also promote change by encouraging the integration of the potentially more advanced perceptual and motor representations (e.g., reasoning about compensation or reversibility) with the potentially less advanced verbally coded representation (e.g., reasoning about a single perceptual dimension on the task objects or reasoning about the transformation).

Note that this view is compatible with the GSA framework (Hostetter & Alibali, 2008), which holds that gesture-speech mismatches occur when some aspects of a speaker's mental representations (i.e., aspects that involve simulations of perception and action) are expressed in gesture and other aspects are expressed in speech. Further, the GSA framework holds that speakers produce gestures when perceptual or action-based aspects of their mental simulations are highly activated (i.e., when activation exceeds the speaker's "gesture threshold"). According to this view, gesture-speech mismatches in conservation indicate that speakers have highly activated mental representations that incorporate perceptual or action-based features of the conservation tasks (hence, they express these ideas in gesture), but they do not express these same mental representations verbally. Evans, Alibali, and McNeil (2001), in their work on conservation understanding in children with Specific Language Impairment, described this situation as a "divergence of verbal expression and embodied knowledge" (p. 309).

From this perspective, transition periods in the acquisition of conservation (and perhaps other concepts as well) can be characterized as periods when learners' embodied, perceptual, and action-based knowledge is rich and highly activated. Further, learners seem not to represent this perception- and action-based knowledge in verbal form (see Goldin-Meadow et al., 1993). For example, children may not have the appropriate words for dimensions such as width or cross-sectional area, or they may not realize that the perceptual features they have encoded map to those words. Alternatively, children's perception- and action-based knowledge may simply be unintegrated with other abstract, verbally coded knowledge that they activate at the same time. For example, children may know, and even state verbally, that both rows of checkers have the same number, but they may also strongly activate their perceptual knowledge that one row is much longer than the other.

Thus, gesture-speech mismatches reflect knowledge that children possess, but that they are either unable to express in speech at that point in time or they have not integrated with their verbally coded knowledge. From this perspective, acquiring knowledge of conservation involves either re-representing perceptual and action-based knowledge in a verbal code or integrating perceptual and action-based knowledge with other verbally coded knowledge.

The Role of Children's Gestures in the Social Construction of Knowledge Change

We have suggested that producing gestures may play a role in children's developing understanding of conservation by highlighting perceptual information, which children then utilize in reasoning about the task. We have further suggested that children produce gesture-speech mismatches when their perception- and action-based knowledge outstrips their ability to express that knowledge in verbal form. It also seems likely that producing gestures may serve a social function. If others are able to detect the knowledge that children express in gestures, it might provide them with information about the leading edge of children's knowledge. These social others could in turn scaffold children's developing understanding.

Do Observers Detect and Interpret Information From Children's Gestures?

The first step in this pathway is for adults or others who interact with children to detect the information that children express in their gestures and, in particular, information that they express uniquely in gesture in gesture-speech mismatches. Indeed, evidence indicates that both adults and children can detect and interpret information that children express in gesture when explaining conservation tasks. A number of studies have shown that observers detect and process information that children express in gestures. These include studies using different types of experimental stimuli, including video clips of children explaining conservation tasks (Church, Kelly, & Lynch, 2000; Goldin-Meadow, Wein, & Chang, 1992; Kelly & Church, 1997, 1998) and "live" presentations of children explaining conservation tasks (Goldin-Meadow & Sandhofer, 1999).

Across studies, two techniques have been used to determine whether observers detect gestured information. One technique is the use of "forced choice" questions. For example, in the Kelly and Church (1998) study, video stimuli of children providing conservation task explanations were created. These video stimuli included: (a) explanations in which gesture and speech provided the same (matching) information (e.g., "That glass is tall and that dish is short" accompanied by a gestures demarcating the tall height of the glass and the short height of the dish), and (b) explanations in which gesture and speech provided different (i.e., mismatching) information (e.g., "that glass is taller than the dish" accompanied by gestures representing the thin width of the glass and the fat width of the dish). Children and adults were shown these stimulus videos and asked to answer questions probing for the information contained in speech and gesture. One question asked for information from speech (e.g., Did the child in the video talk about the heights of the containers?). A "yes" response indicated that information from speech was detected. Another question probed for information expressed in gesture in mismatches (e.g., Did the child talk about the widths of the containers?). A "yes" response indicated that the information expressed uniquely in gestures was detected, and a "no" response indicated failure to detect the information expressed uniquely in gestures. In addition, two "false-positive questions" were asked (e.g., Did the child talk about the color of the containers?), so that authentic detection could be distinguished from response bias. Kelly and Church (1998) found that "yes" responses to questions probing for spoken and gestured information were significantly more frequent than "yes" responses to false-positive questions.

A second technique for determining whether observers detect gestured information requires observers to reiterate or restate what they heard in the video (Church et al., 2000; Goldin-Meadow et al., 1992; Kelly & Church, 1998). Note that this question asks observers to repeat what they heard in the video stimulus, slanting the bias toward paying attention only to the stimulus speech. When the reiteration also includes information expressed in mismatching gesture, it suggests that the observer's attention to the stimulus gesture is indeed compelling. For example, after viewing each of the video stimuli described above, the observer would be asked to repeat what the child in the video said. Responses were classified as: (a) repeating what the child said, (b) adding additional information that was traceable to the child's gestures, or (c) adding additional information that was not traceable to the child's gestures. Thus, when an observer viewed the gesture-speech mismatch described above and said, "The child said that the dish was shorter and fatter than the glass," the information about the width of the glass would be classified as "additional information traceable to the child's gestures," thus indicating detection of information from the child's gestures.

Several studies using these techniques have shown that gesture affects how speech is processed. For example, Kelly and Church (1997, 1998) showed that both children and adults were better able to recall conservation explanations that children expressed in redundant speech and gestures than explanations that children expressed in spoken explanations accompanied by mismatching gestures. Thus, gestures that express different information from speech appear to interfere with observers' processing of speech.

What about the information that children express uniquely in gesture in gesture-speech mismatches? Observers do detect this information, and it enriches their understanding of speakers' verbal communication. As described above, when asked to view video clips of children explaining conservation tasks, both child and adult observers often detected and interpreted the information that children expressed uniquely in gestures (Church et al., 2000; Goldin-Meadow et al., 1992; Kelly & Church, 1998), and they sometimes incorporated the information children expressed in gestures into their interpretations of the children's speech (Goldin-Meadow et al., 1992; Kelly & Church, 1997), as in the example described above. Thus, observers sometimes "credit" children with the embodied knowledge that they express in their gestures.

Others May Alter Their Interactions With Children Based on Information From Gesture

Thus, not only do observers detect children's gestures, they also incorporate information from children's gestures into their interpretations of what children said and what they understand. It is possible that observers might then alter their interactions with children on the basis of this gestured information—perhaps providing crucial information to help children progress to a more advanced understanding. Indeed, one might predict that it might be particularly valuable for children to receive input that helps them bridge from their embodied, perception- and action-based understandings of the conservation tasks to more abstract, logic-based forms of reasoning about the tasks. We know of no data that directly address this possibility for conservation; however, for studies of children learning mathematical equations, evidence suggests that adults do alter their input to children on the basis of the children's gestures (Goldin-Meadow & Singer, 2003).

Given that people do detect and interpret the gestures that others produce, one might ask whether gestural input that accompanies speech in instructional settings might also have an effect on learning. We turn next to studies that have focused on gesture in instruction about conservation.

Children Learn About Conservation From Instruction That Includes Both Gesture and Speech

Teachers routinely use gestures in their instruction (e.g., Alibali, Nathan, & Fujimori, 2010), and growing evidence suggests that instructors' gestures make a difference for their students' learning. This holds true for many concepts and many age groups, including elementary school students learning about equations (Singer & Goldin-Meadow, 2005), kindergarten students learning about symmetry (Valenzeno, Alibali, & Klatzky, 2003), and, of course, kindergarten and first-grade students learning about conservation.

Two studies to date have investigated whether children improve in their conservation understanding in response to instruction that includes gesture. One study utilized video instruction and the other utilized "live" instruction. In the video study (Church, Ayman-Nolley, & Mahootian, 2004), children received one of two videotaped lessons: one with speech alone and one with gestures that expressed the same information as speech. For example, the instructor said, "This container has the same amount of water as the glass because the glass may be taller, but it is also skinnier than the other container." In the speech-plus-gesture lesson, the instructor produced gestures accompanying this explanation that conveyed that the glass was both tall and skinny. Children who viewed the speech-plus-gesture lesson were twice as likely to improve on the conservation task compared with children who watched the speech-only lesson.

In the study of "live" instruction (Ping & Goldin-Meadow, 2008), the lessons utilized either concrete objects or no objects, and the information was provided in either speech alone or speech with gesture. Like the video study, this study revealed that children were more likely to learn from instruction that included speech-plus-gesture instruction than from speech-only instruction. They were more likely to provide correct judgments after instruction and also more likely to adopt the conservation explanation that was provided during instruction (compensation). Moreover, these learning benefits held, both when the instruction referred to concrete objects and when the instruction was provided with no objects present. In fact, children who received the lesson without objects but with gesture showed the deepest learning, often adding new types of correct conservation explanations that had not been directly taught in the lesson. Ping and Goldin-Meadow (2008) speculated that when gestures are not grounded on objects, they might encourage learners to construct more abstract representations of the problem, which may transfer more readily to other sorts of problems.

How Do Instructional Gestures Facilitate Learning?

From an embodied cognition perspective, there are several possible mechanisms by which instructional gestures may facilitate learning. Gestural input that grounds the instructor's speech in the physical environment, like the gestures used in the Church et al. (2004) study and those used in the objects-present condition in the Ping and Goldin-Meadow (2008) study, may help children better understand the instructor's speech, which may yield benefits for learning. In addition, the instructor's gestures may help learners construct appropriate simulations of the task context even when objects are not present, and these simulations may in turn support learning and performance (see Alibali & Hostetter, 2011).

CONCLUSION

In this chapter, we have taken an embodied perspective on the development of understanding of Piagetian conservation, with a focus on the embodied knowledge that is manifested in children's spontaneous gestures in their explanations of conservation tasks. We have argued that gestures are a form of action that may be involved in mechanisms of change in the development of conservation understanding. We considered two possible types of mechanisms: (a) mechanisms that involve co-activating and integrating embodied and verbally coded knowledge, and (b) mechanisms that involve others detecting the embodied knowledge that children express in gestures and offering input tailored to that knowledge. Such input may help children re-represent their existing embodied knowledge in verbal form or may introduce information that is crucial for progress to a more advanced, logic-based understanding of conservation. We have focused here on change in conservation understanding, but we believe that cognitive change in other domains may follow the same process.

One of the central claims of Piaget's theory is that abstract concepts have their roots in sensorimotor experiences (e.g., Piaget, 1936/1952). In this chapter, we have presented evidence suggesting that the path from concrete, sensorimotor understanding to abstract understanding is manifested in a shift from perception- and action-based representations that are readily expressed in gesture to more abstract representations that are readily expressed in speech. Gestures are iconic and deictic representations of perceptions and actions in the world, and yet gestures do not actually manipulate or act on the environment. Thus, gesture can serve as a sort of "segue" or bridge between actual actions and perceptions and abstract, verbally coded representations.

The evolutionary history of gesture as a communication system tells a similar story. Many investigators believe that language emerged gradually from an action-based gestural system of communication (Armstrong & Wilcox, 2007). The claim is that the gestural system evolved out of hands-on actions and then evolved into a language system, which supplanted the gesture system. From this perspective, it is not surprising that gesture reflects embodied understanding, not only for conservation reasoning but also for many other concepts.

In summary, our embodied perspective on the development of conservation understanding hearkens back to the core ideas of Piagetian constructivism and, in particular, the idea that sensorimotor knowledge provides a foundation for more abstract, logic-based understanding. We have argued that children's gestures provide a window into their embodied knowledge of conservation. Further, we have suggested that gesture serves as an important bridge between perception- and action-based knowledge and more abstract knowledge that can be expressed in words. Thus, studies of gesture in children's conservation explanations provide a revealing glimpse into change processes that involve the gradual abstraction of embodied knowledge.

REFERENCES

Alibali, M. W., & Hostetter, A. B. (2011). Mimicry and simulation in gesture comprehension. (Commentary on P. Niedenthal, M. Maringer, M. Mermillod, & U. Hess, *The Simulation of Smiles (SIMS) model: Embodied simulation and the meaning of facial expression.*) *Behavioral and Brain Sciences, 33,* 433–434.

Alibali, M. W., & Kita, S. (2010). Gesture highlights perceptually present information for speakers. *Gesture, 10*(1), 3–28.

Alibali, M. W., Kita, S., & Young, A. (2000). Gesture and the process of speech production: We think, therefore we gesture. *Language and Cognitive Processes, 15,* 593–613.

Alibali, M. W., & Nathan, M. J. (2012). Embodiment in mathematics teaching and learning: Evidence from students' and teachers' gestures. *Journal of the Learning Sciences, 21*(2), 247–286.

Alibali, M. W., Nathan, M. J., & Fujimori, Y. (2010). Gesture in the mathematics classroom: What's the point? In N. Stein & S. Raudenbush (Eds.), *Developmental cognitive science goes to school.* New York, NY: Routledge.

Alibali, M. W., Spencer, R. C., Knox, L., & Kita, S. (2011). Spontaneous gestures influence strategy choices in problem solving. *Psychological Science, 22*(9), 1138–1144.

Armstrong, D. F., & Wilcox, S. (2007). *The gestural origin of language.* New York, NY: Oxford University Press.

Barsalou, L. W. (2008). Grounded cognition. *Annual Review of Psychology, 59,* 617–645.

Ben-Zeev, T., & Star, J. R. (2001). Spurious correlations in mathematical reasoning. *Cognition and Instruction, 19,* 253–275.

Church, R. B. (1999). Using gesture and speech to capture transitions in learning. *Cognitive Development, 14,* 313–342.

Church, R. B., Ayman-Nolley, S., & Mahootian, S. (2004). The role of gesture in bilingual education: Does gesture enhance learning? *International Journal of Bilingual Education and Bilingualism, 7,* 303–319.

Church, R. B., & Goldin-Meadow, S. (1986). The mismatch between gesture and speech as an index of transitional knowledge. *Cognition, 23,* 43–71.

Church, R. B., Kelly, S. D., & Lynch, K. (2000). Immediate memory for mismatched speech and representational gesture across development. *Journal of Nonverbal Behavior, 24,* 151–174.

Evans, J. L., Alibali, M. W., & McNeil, N. M. (2001). Divergence of embodied knowledge and verbal expression: Evidence from gesture and speech in children with Specific Language Impairment. *Language and Cognitive Processes, 16,* 309–331.

Gallese, V., & Lakoff, G. (2005). The brain's concepts: The role of the sensory-motor system in conceptual knowledge. *Cognitive Neuropsychology, 22*(3/4), 455–479.

Gelman, R. (1969). Conservation acquisition: A problem of learning to attend to relevant attributes. *Journal of Experimental Child Psychology, 7,* 167–187.

Gibbs, R. W. (2006). *Embodiment and cognitive science.* Cambridge, England: Cambridge University Press.

Glenberg, A. M. (2010). Embodiment as a unifying perspective for psychology. *Wiley Interdisciplinary Reviews: Cognitive Science, 1*(4), 586–596.

Goldin-Meadow, S., Alibali, M. W., & Church, R. B. (1993). Transitions in concept acquisition: Using the hand to read the mind. *Psychological Review, 100,* 279–297.

Goldin-Meadow, S., & Sandhofer, C. M. (1999). Gesture conveys substantive information to ordinary listeners. *Developmental Science, 2,* 67–74.

Goldin-Meadow, S., & Singer, M. A. (2003). From children's hands to adults' ears: Gesture's role in the learning process. *Developmental Psychology, 39,* 509–520.

Goldin-Meadow, S., Wein, D., & Chang, C. (1992). Assessing knowledge through gesture: Using children's hands to read their minds. *Cognition and Instruction, 9*, 201–219.

Hostetter, A. B., & Alibali, M. W. (2008). Visible embodiment: Gestures as simulated action. *Psychonomic Bulletin and Review, 15*, 495–514.

Houdé, O., Pineau, A., Leroux, G., Poirel, N., Perchey, G., Lanoë, C., & Mazoyer, B. (2011). Functional magnetic resonance imaging study of Piaget's conservation-of-number task in preschool and school-age children: A neo-Piagetian approach. *Journal of Experimental Child Psychology, 110*(3), 332–346.

Irvine, J. T. (1978). Wolof "magical thinking": Culture and conservation revisited. *Journal of Cross-Cultural Psychology, 9*(3), 300–310.

Jeannerod, M. (2001). Neural simulation of action: A unifying mechanism for motor cognition. *NeuroImage, 14*, S103–S109.

Kelly, S. D., & Church, R. B. (1997). Can children detect conceptual information conveyed through other children's nonverbal behaviors? *Cognition and Instruction, 15*, 107–134.

Kelly, S. D., & Church, R. B. (1998). A comparison between children's and adults' ability to detect conceptual information conveyed through representational gestures. *Child Development, 69*, 85–93.

Klahr, D., & Wallace, J. G. (1973). The role of quantification operators in the development of conservation of quantity. *Cognitive Psychology, 4*(3), 301–327.

Klahr, D., & Wallace, J. G. (1976). *Cognitive development: An information processing view.* Hillsdale, NJ: Lawrence Erlbaum Associates.

Lakoff, G., & Johnson, M. (1980). *Metaphors we live by.* Chicago, IL: University of Chicago Press.

Lakoff, G., & Núñez, R. (2001). *Where mathematics comes from: How the embodied mind brings mathematics into being.* New York, NY: Basic Books.

Landy, D., & Goldstone, R. L. (2007). How abstract is symbolic thought? *Journal of Experimental Psychology: Learning, Memory, and Cognition, 33*, 720–733.

Mainela-Arnold, E., Evans, J. L., & Alibali, M. W. (2006). Understanding conservation delays in children with Specific Language Impairment: Task representations revealed in speech and gesture. *Journal of Speech, Language, and Hearing Research, 49*, 1267–1279.

McGarrigle, J., & Donaldson, M. (1974). Conservation accidents. *Cognition, 2*, 341–350.

McNeill, D. (2005). *Gesture and thought.* Chicago, IL: University of Chicago Press.

Núñez, R. (2005). Do real numbers really move? Language, thought, and gesture: The embodied cognitive foundations of mathematics. In F. Iida, R. Pfeifer, L. Steels, & Y. Kuniyoshi (Eds.), *Embodied artificial intelligence* (pp. 54–73). Berlin, Germany: Springer-Verlag.

Núñez, R., Edwards, L. D., & Matos, J. F. (1999). Embodied cognition as grounding for situatedness and context in mathematics education. *Educational Studies in Mathematics, 39*, 45–65.

Özçaliskan, S., & Goldin-Meadow, S. (2009). When gesture-speech combinations do and do not index linguistic change. *Language and Cognitive Processes, 24*(2), 190–217.

Perry, M., Church, R. B., & Goldin-Meadow, S. (1988). Transitional knowledge in the acquisition of concepts. *Cognitive Development, 3*, 359–400.

Piaget, J. (1952). *The origins of intelligence in children.* New York, NY: International University Press. (Original work published 1936)

Piaget, J. (1952). *The child's conception of number.* London, England: Routledge and Kegan Paul. (Original work published 1941)

Pine, K. J., Lufkin, N., & Messer, D. (2004). More gestures than answers: Children learning about balance. *Developmental Psychology, 40*(6), 1059–1067.

Ping, R., & Goldin-Meadow, S. (2008). Hands in the air: Using ungrounded iconic gestures to teach children conservation of quantity. *Developmental Psychology, 44*, 1277–1287.

Ping, R., & Goldin-Meadow, S. (2010). Gesturing saves cognitive resources when talking about nonpresent objects. *Cognitive Science, 34*, 602–619.

Price-Williams, D., Gordon, W., & Ramirez, M. (1969). Skill and conservation: A study of pottery-making children. *Developmental Psychology, 1*(6), 769.

Rains, J. R., Kelly, C. A., & Durham, R. L. (2008). The evolution of the importance of multi-sensory teaching techniques in elementary mathematics: Theory and practice. *Journal of Theory and Practice in Education, 4*(2), 239–252.

Roth, W.-M. (2002). Gestures: Their role in teaching and learning. *Review of Educational Research, 71*, 365–392.

Schliemann, A. D., & Carraher, D. W. (2002). The evolution of mathematical reasoning: Everyday versus idealized understandings. *Developmental Review, 22*, 242–266.

Schultz, T. R. (1998). A computational analysis of conservation. *Developmental Science, 1*, 103–126.

Siegal, M., & Waters, L. J. (1988). Misleading children: Causal attributions for inconsistency under repeated questioning. *Journal of Experimental Child Psychology, 45*, 438–456.

Siegler, R. S., & Robinson, M. (1982). The development of numerical understandings. In H. W. Reese & L. P. Lipsitt (Eds.), *Advances in child development and behavior* (Vol. 16, pp. 242–312). New York, NY: Academic Press.

Singer, M. A., & Goldin-Meadow, S. (2005). Children learn when their teacher's gestures and speech differ. *Psychological Science, 16*, 85–89.

Valenzeno, L., Alibali, M. W., & Klatzky, R. L. (2003). Teachers' gestures facilitate students' learning: A lesson in symmetry. *Contemporary Educational Psychology, 28*, 187–204.

van der Maas, H. L. J., & Molenaar, P. C. M. (1992). Stagewise cognitive development: An application of catastrophe theory. *Psychological Review, 99*(3), 395–417.

van Geert, P. (1998). A dynamic systems model of basic developmental mechanisms: Piaget, Vygotsky and beyond. *Psychological Review, 105*(4), 634–677.

Wilson, M. (2002). Six views of embodied cognition. *Psychonomic Bulletin & Review, 9*, 625–636.

Zhang, Q., Shi, J., Fan, Y., Liu, T., Luo, Y., Sang, H., & Shen, M. (2008). An event-related brain potential study of children's conservation. *Neuroscience Letters, 431*(1), 17–20.

CHAPTER 3

GESTURE'S ROLE IN LEARNING ARITHMETIC

Susan Goldin-Meadow
Susan C. Levine
Steven Jacobs
The University of Chicago

Mathematical concepts are traditionally viewed as abstract and formal. However, recent research suggests that there is an embodied component to these concepts—that notions like infinity, continuity, and number are metaphorical and grounded in our experience of the world (Lakoff & Núñez, 2000). The evidence for this view has largely come from analyses of mathematical language. But the spontaneous gestures that speakers produce when they talk about math have also been used to argue that mathematical concepts are metaphorical and embodied (Núñez, 2008). For example, gesture can reveal that a speaker construes a problem dynamically even when the speaker's words focus exclusively on the problem's static notation (e.g., Marghetis & Núñez, 2010).

Perhaps, then, it is not surprising to find that children often gesture when talking about arithmetic, the foundational branch of mathematics. For example, toddlers routinely point when counting (e.g., Fuson, 1988; Gelman & Gallistel, 1978; Graham, 1999; Saxe, 1977), and school-age children often

Emerging Perspectives on Gesture and Embodiment in Mathematics, pages 51–72

move their hands when explaining their solutions to mathematical equivalence problems (Perry, Church, & Goldin-Meadow, 1988). Gesture enables young speakers to use their bodies to instantiate number and simple operations on those numbers; for example, producing a sweeping gesture under a string of numbers to indicate that they should be added together.

But learners are not the only ones who gesture when talking about arithmetic. Parents frequently move their hands when counting sets with their young children (Suriyakham, 2007), and instructors frequently move their hands when teaching arithmetic to individual students (Goldin-Meadow, Kim, & Singer, 1999) and in their classrooms (Flevares & Perry, 2001). Some gestures are simple deictic points (e.g., grounding a spoken number with a point to a numeral written on the board); others are more complex and representational (e.g., demonstrating division by two by holding up two hands in front of the body and then moving the hands away from each other).

Here we describe the pedagogical role that gesture can play in teaching and learning arithmetic. We focus on two arithmetic skills—counting and simple arithmetic operations. We first show that gesture can serve as a window into children's early arithmetic understanding; gesture provides an additional and often overlooked modality for discovering what children know about arithmetic. We then show that gesture can serve as an instructional tool in arithmetic lessons, both the gestures that learners produce themselves and the gestures they see their teachers produce. Gesture can thus not only be used to discover what children know about arithmetic but may also change what they know.

GESTURE REFLECTS WHAT CHILDREN KNOW ABOUT ARITHMETIC

Counting

Toddlers routinely use their fingers when counting, and their hand movements may not only enable children to count more accurately (e.g., Gelman & Gallistel, 1978) but may also reveal aspects of their early counting skills and understanding that are not apparent in their talk (e.g., Gelman & Gallistel, 1978; Saxe, 1977). Gelman and Gallistel (1978) found that 2-year-olds pointed on at least some trials when asked to count. Saxe (1977) examined the counting behaviors of 3- and 4-year-old children who were asked to compare and reproduce sets containing from 4 to 9 objects. At age 3, children either did not gesture or failed to coordinate the points they did produce with their number words. By age 4, children almost always produced pointing gestures while counting, and they did so successfully, in a one-to-one manner, on the majority of trials (63%). Broadly consistent with Saxe's findings, in a longitudinal study of parent-child naturalistic interactions, Suriyakham (2007) found that at 30 months, children produced more counting utterances without pointing gestures than with them; by 38 months, they had increased their pointing gestures, producing roughly equivalent numbers of counting utterances with and without gesture. Together these studies indicate that pointing while counting is a developmental advance. Although Suriyakham did not compare the accuracy of counts with and without gesture, other research findings suggest that counts accompanied by gesture are more accurate than counts without gesture (e.g., Alibali & DiRusso, 1999; Potter & Levy, 1968; Saxe & Kaplan, 1981).

Other studies have examined the specificity of children's pointing gestures, asking whether children point more when counting larger versus smaller sets and whether they point more when counting the elements in a set versus labeling the cardinal value of the set. Graham (1999) asked 2-, 3-, and 4-year-old children to count sets of two, four, and six object arrays. Not surprisingly, children use small numbers and count small sets before learning to use large numbers and count large sets. If children gesture primarily when the counting problem is hard, we might expect them to gesture more on large arrays (four and six objects) than on small arrays (two objects), which can be enumerated without counting (i.e., on sets than can be subitized; Gelman & Gallistel, 1978; Saxe & Kaplan, 1981; Levine, Suriyakham, Rowe, Huttenlocher, & Gunderson, 2011). The 4-year-olds did just that (the 2- and 3-year-olds seemed to be challenged by all three arrays and gestured equally often on all of them). Thus, by age 4, children seem to be recruiting gesture when the counting task becomes difficult, indicating that they are not applying this behavior indiscriminately but rather to help them obtain the correct answer.

Suriyakham (2007) asked whether 30- and 38-month-olds use gesture differently when counting items in a set (e.g., "one, two, three") versus when labeling the cardinal value of items in the set (e.g., "I have three trucks"). At 30 months, children produced gesture equally often in counting versus cardinal contexts. However, at 38 months, they produced gesture more often in counting than in cardinal contexts, thus clearly differentiating between the two, an important step in the acquisition of the cardinal meaning of number words (e.g., Le Corre, Van de Walle, Brannon, & Carey, 2006; Wynn, 1990). Given prior research indicating that gesture often predicts and promotes verbal understanding (Goldin-Meadow, 2003), these findings raise the possibility that the children who use gesture differentially when counting, compared with when labeling set size, might be just those children who are ready to advance in their understanding of cardinal number.

Even when judging others' explanations of counting, children seem to rely on gesture. When children notice a speaker's counting words, they also notice that speaker's counting gestures but not vice versa. Graham (1999) asked very young children to "help" a puppet learn to count. Half the time the puppet counted correctly, but the other half the time the puppet added an extra number (e.g., the puppet would say "one, two three" while counting two objects). In addition, when the puppet made these counting errors, he either produced the same number of pointing gestures as number words (three in this example), a larger or smaller number of pointing gestures (four or two pointing gestures), or no pointing gestures at all. The child's job was to tell the puppet whether his counting was correct and, if incorrect, to explain why the puppet was wrong. The interesting result from the point of view of this discussion is whether children made reference to the puppet's number words (speech only), the puppet's points (gesture only), or both (gesture+speech) in their explanations. Two-year-olds did not refer to either gesture or speech, 3-year-olds referred to gesture but not speech (gesture only), and 4-year-olds referred to both gesture and speech (gesture+speech). Few children across all three ages referred to the puppet's counting words without also referring to the puppet's counting gestures. In other words, when they noticed the puppet's words, they also noticed his gestures but not necessarily vice versa.

In summary, pointing while counting emerges early and, at first, is not perfectly coordinated with the number of objects in a set or with the number words the child produces. By age 3, children produce more pointing gestures while counting than while labeling the cardinality of sets (Suriyakam, 2007). By age 4, children selectively produce pointing gestures while counting—they point more when counting larger sets than smaller sets perhaps because they realize that pointing increases the accuracy of their count. Moreover, children notice the accuracy of others' points while counting even when they do not notice the accuracy of their own counting words. These findings suggest that pointing gestures that accompany counting reflect what the child knows about counting. The findings also indicate the fundamental and persistent embodied nature of our mathematical representations.

Going one step further, researchers have suggested that pointing gestures can also play a role in the development of counting. For example, pointing while counting may help children partition items into those counted versus not counted (Potter & Levy, 1968). Moreover, pointing while counting has been hypothesized to play a supportive role in the development of the fundamental counting principles, including one-to-one correspondence (i.e., that each number in the count sequence is paired with an item in the set; Gelman & Gallistel, 1978), the stable order principle (i.e., that the count words need to be uttered in a fixed order; Wiese, 2003), and the cardinal principle (i.e., that the last number in the count sequence indicates set size; Fayol & Seron, 2005).

Butterworth (1999) uses the fact that impaired finger representation (finger agnosia) is associated with imprecise numerical representations as support for the claim that fingers play a role in the development of counting. Finger gnosia (i.e., the ability to recognize fingers) is typically assessed by asking children or adults to indicate which finger has been touched or by asking them to determine whether two fingers were touched simultaneously or successively without visual access to this information (e.g., Fayol, Barrouillet, & Marinthe, 1998; Noël, 2005). Children's performance on a finger gnosis test at age 5 predicted their math skills (but not their reading skills) at age 8 and did a better job of predicting math skills than performance on intellectual tasks (Fayol et al., 1998; see also Marinthe, Fayol, & Barrouillet, 2001). The tight relation between finger gnosia and performance on math tasks could be due to the proximity of the neural areas that subserve these functions (both the parietal cortex and the precentral gyrus are involved in finger and number representations; Noël, 2005). Alternatively, fingers could be serving a functional role in numerical representations—under this view, poor finger gnosis impairs children's ability to use their fingers to count, which in turn impairs their ability to develop strong numerical skills (Gracia-Bafalluy & Noël, 2008). Finger-differentiation training has, in fact, been found to improve numerical skills in first-grade students who scored low on a finger gnosis test (Gracia-Bafalluy & Noël, 2008).[1]

A related but somewhat different research approach has asked whether preschool children are better at using and understanding finger representations for set size than using number word representations for set size (Crollen, Seron, & Noël, 2011; Nicoladis, Pika, & Marantette, 2010) perhaps because finger representations provide a more iconic representation (Wiese, 2003), which could facilitate arriving at the correct count. Nicoladis et al. (2010) tested this possibility in two studies: (a) they asked children to put a number of toys into a box, which was identified by a number word or a hand shape; and (b) they presented children with a set of toys and asked the children to tell them how many toys were in the set, using either a number word or a hand shape. In neither study was gesture found to have an advantage over number words. Although these studies do not show a gesture advantage, it is important to note that the tasks used are different from those used in the studies reported earlier, which examine children's spontaneous reliance on fingers during counting tasks. This difference suggests that children's spontaneous use of gesture while counting may reflect implicit knowledge about the connection between counting and cardinality rather than explicit knowledge that can be used to answer a "how many" question. As suggested by both Nicoladis et al. (2010) and Crollen et al. (2011), it is also possible that spontaneous use of gesture while counting goes beyond reflecting children's implicit numerical knowledge to help them develop explicit numerical knowledge.

Taken together, these findings raise the possibility that early pointing gestures not only reflect what children know about early number but also play a role in creating that knowledge. To test this hypothesis, we need to experimentally manipulate gesture, a point to which we return in a later section.

Arithmetic Operations

Early Calculation

Levine, Jordan, and Huttenlocher (1992) examined how 4½- to 7-year-old children used their fingers while solving addition and subtraction problems presented in three different formats: (a) number fact problems—for example, "How much is *m* and *n*?" How much is *m* take away *n*?"; (b) story problems—for example, "Mike had *m* balls; he got *n* more; how many balls did he have all together?" "Kim had *m* crayons; she lost *n*; how many crayons did she have left?"; and (c) nonverbal number problems—for example, a set of disks was placed on a mat and then covered with a box; another set of disks was placed on the mat and the disks were slid under the box one by one through a small opening on the side of the box; the child was asked to lay out the number of disks that were under the box on the mat. A comparable procedure was used for the subtraction problems. For each of the calculation problem types, finger strategies were coded if children were observed counting on their fingers or if they held up their fingers for any term in a problem without overtly counting them.

Levine and colleagues (1992) hypothesized that using fingers when solving calculation problems could serve the same function as manipulatives; that is, it could help children carry out a counting strategy when they cannot retrieve the correct answer and also reduce the working memory load of mentally imagining the to-be-counted objects. They thus predicted that children would use fingers most often on the number fact problems that made no reference to objects, next most often on the story problems that referred to objects that were not physically present, and least often on the nonverbal number problems that referred to physically present objects. They found this ordering for finger strategies in the two oldest age groups, children ages 5 years 6 months to 5 years 11 months and children ages 6 years 0 months to 6 years 5 months (children below 5½ years of age rarely used their fingers when calculating). Moreover, for each problem type, the older age groups were more accurate when they used their fingers than when they did not (90% versus 78% for nonverbal problems, 83% versus 66% for story problems, 75% versus 54% for number fact problems). In fact, accuracy on the two harder types of problems (story problems and number fact problems) when fingers were used (83%, 75%) was as good as accuracy on the easiest problems (nonverbal problems) without gesture (78%). These findings suggest that using fingers on story problems and number fact problems can substitute for the presence of concrete objects and thus raise performance levels.

Subsequent studies showed a difference in how often middle- versus low-income kindergarten and first-grade children used their fingers when asked to solve calculation problems in the same three formats. Middle-income children used fingers more often than low-income children in kindergarten, and low-income children used fingers more often than middle-income children in first grade (e.g., Jordan, Huttenlocher, & Levine, 1992; Jordan, Levine, & Huttenlocher, 1994). Following up on these studies, Jordan, Kaplan, Ramineni, and Locuniak (2008) studied middle- and low-socioeconomic status (SES) children from the fall of kindergarten through the spring of second grade and examined how these children used their fingers to solve number fact problems and whether finger use correlated with problem-solving success over developmental time. Finger use decreased between kindergarten and second grade for middle-SES children but increased over this period for low-SES children. Moreover, consistent with the Levine et al. (1992) findings, using finger strategies in kindergarten was a strong, positive predictor of calculation success in second grade. By the end of second grade, finger use was a negative predictor of children's calculation success.

Siegler and colleagues (e.g., Siegler & Jenkins, 1989; see also Beller & Bender, 2011) have shown that how children use their fingers when solving calculation problems at the least reflects, and may even scaffold, their calculation knowledge and fluency. Initially, children use the "count all" strategy, holding up the number of fingers represented by each addend of the problem (e.g., holding up two fingers, then another four, for the problem 2 + 4) and then counting all of the raised fingers (i.e., "one, two, three, four, five, six"). They then shift to the "count on" strategy, either counting on from the first addend (i.e., counting on from the 2, "three, four, five, six") or using the more efficient min strategy and counting on from the largest addend (i.e., counting on from the 4, "five, six").

In summary, young children use their fingers to represent the cardinal value of sets and to count when solving calculation problems, particularly when other manipulatives are not available. The way in which they use their fingers changes over the course of development depending on their knowledge and the ease with which they can retrieve answers from memory. At early ages, children who use their fingers perform better on calculation problems than children who do not. In contrast, at later ages, children who use their fingers perform worse on calculation problems than children who do not most likely because the non-gesturers at this age are able to mentally represent the cardinal value of sets and retrieve the answers to calculation problems without counting (although the non-gesturers could also be recognizing the quantity without mentally representing the size of the sets; that is, they may simply have memorized the math facts and be using retrieval strategies to answer the problems).

Mathematical Equivalence

One basic concept that children need to master, one that is surprisingly difficult, is the meaning of the equal sign. Children as old as 9 or 10 years often think that the equal sign is merely an instruction to add up all of the numbers in a problem and fill in the blank. This interpretation works well if the problem is a simple one laid out in a traditional format (as most problems in U.S. textbooks are; McNeil et al., 2006) (e.g., $7 + 6 + 4 = \_\_\_$). But the same children who can solve this problem correctly fail if the blank is situated on the left side of the equation (e.g., $7 + 6 + \_\_\_ = 17$) or if an additional number is placed on the right side of the equation (e.g., $7 + 6 + 4 = 7 + \_\_\_$), thus failing to appreciate the meaning of the equal sign (that the quantities on the two sides of the equation must add up to equivalent numbers). When children fail to solve a problem like $7 + 6 + 4 = 7 + \_\_\_$ correctly, they typically either add up all of the numbers in the problem and put, in this case, 24 in the blank, or they add up all of the numbers on the left side of the equal side and put 17 in the blank. When asked to explain their incorrect solutions to these problems, children typically say what they did (e.g., "I added the 7, the 6, the 4, and the 7 and got 24," the add-all-numbers strategy, or "I added the 7, the 6, and 4 and got 17," the add-to-equal-sign strategy).

However, most children also gesture as they explain their solutions in speech. Some convey the same information in their gestures as they convey in speech (e.g., they say, "I added 7, 6, and 4 and got 17" while pointing at the 7, 6, and 4, and then the blank, the add-to-equal-sign strategy in both speech and gesture). These children produce what have been called "gesture-speech matches" (Church & Goldin-Meadow, 1986; Goldin-Meadow, 2003). However, other children convey different information in their gestures than they convey in their speech—they produce "gesture-speech mismatches" (Church & Goldin-Meadow, 1986; Goldin-Meadow, 2003). For example, a child might say, "I added 7, 6, and 4 and got 17" (an add-to-equal-sign strategy) while pointing at all of the numbers in the problem, at the 7, 6, and 4 on the left side of the equation and the 7 on the right side (an add-all-numbers strategy), thus making it clear that she did, at some level, know that the 7 on the right side of the equation was there and might be important. Note that this second child seems to have an understanding (however implicit) of two pieces of information: (a) there are two distinct sides to the equation (reflected in the add-to-equal-sign strategy the child conveyed in speech), and (b) there is an additional addend on the right side of the equation (reflected in the add-all-numbers strategy conveyed in gesture). These two pieces of information are not yet integrated into a single framework but will have to be if the child is to solve the problem correctly.

Children who produce mismatches in their explanations of a task have information relevant to solving the task at their fingertips and could, as a result, be on the cusp of learning the task. If so, they may be particularly susceptible to instruction. Perry, Church, and Goldin-Meadow (1988) gave 9- and 10-year-old children instruction on problems of the $4 + 5 + 3 = \_\_\_ + 3$ variety. Prior to instruction, all of the children solved the problems incorrectly, and all of their spoken explanations were incorrect. However, the children differed with respect to their gestures: Some produced gestures that did not match their speech, whereas others produced only matching gestures. After the instruction period, the children were given a second test to see how much they learned. Children who produced mismatches prior to instruction learned more than children who produced no mismatches (Perry et al., 1988).

When children's responses to problems of this type are charted over the course of instruction, we can see a child systematically progress through three periods characterized by the relation between gesture and speech: (a) the child first produces the same strategy in both speech and gesture and that strategy is incorrect; (b) the child then produces two different strategies, one in speech and a different one in gesture; both strategies may be incorrect, or one may be correct and the other incorrect (if there is a correct strategy, it is typically produced in gesture and not speech); and (c) the child again produces the same strategy in both speech and gesture, but now that strategy is correct (Alibali & Goldin-Meadow, 1993). Gesture, when taken in relation to speech, signals that the child is ready to take the next step in learning about mathematical equivalence. Interestingly, when a child fails to pass through step (b) and goes directly from step (a) to step (c), the child's understanding of mathematical equivalence is likely to be fragile—the child is unable to generalize the knowledge gained during instruction and does not retain the knowledge on a follow-up test (Alibali & Goldin-Meadow, 1993).

The gestures that children produce when talking about mathematical equivalence thus provide insight into ideas that have not yet appeared in their speech. Gesture offers a second window into children's knowledge of arithmetic operations, a window that can tell us who is ready to learn that operation. If teachers not only listen to what their students say, but are also sensitive to the information they convey in their gestures (as they often are; Goldin-Meadow & Singer, 2003), they can use this information in planning their lessons and thus promote student learning.

GESTURE CAN CHANGE WHAT CHILDREN KNOW ABOUT ARITHMETIC

We have seen that gesture can reflect what children know about arithmetic and, importantly, can provide insight into knowledge not evident in their speech. But recent research has shown that gesture can go beyond reflecting what children know about arithmetic to play a role in changing what they know. Gesture can play a role in knowledge change in two ways: The gestures children see and the gestures children do can influence whether they will learn a particular arithmetic concept.

The Gestures Children See

The Gestures Parents Produce During Counting

Many studies have noted the presence of pointing gestures when sets of objects are being counted. As a set of objects is being counted, adults as well as young children frequently point to each object in turn as they recite the count list (e.g., Fuson & Hall, 1983, Gelman & Meck, 1983; Graham, 1999; Saxe & Kaplan, 1981; Schaeffer, Eggleston, & Scott, 1974). However, we do not know whether exposure to more gesture in the input, either during counting or when set sizes are labeled, affects children's ability to count or their understanding of the cardinal principle.

Suriyakham (2007) examined parents' use of gesture when they counted and labeled set sizes for their 30- and 38-month-old children. Parents, like their 38-month-old children (described earlier), use gestures significantly more often when counting than when labeling set sizes. Moreover, the kinds of gestures they produce in these contexts differ. When counting, they most often sequentially point to each of the items in the set. In contrast, when gestures were produced in the context of a cardinality utterance, the most common gesture they produced was a single point to the set, and the second most common gesture in this context was an iconic hand shape (e.g., holding up two fingers to indicate "two").

Suriyakham (2007) next examined whether children whose parents produced more gestures when counting and labeling set sizes had stronger numerical knowledge than children whose parents produced fewer gestures when counting and labeling set sizes. She found that children whose parents were in the high gesture group said more number words than those whose parents were in the low gesture group; they also performed better on the Point-to-X task (in which children are presented with two sets, for example, one with three items and one with four items, and are asked to point to four items). Importantly, the two groups of parents did not differ significantly in terms of the frequency of their number talk, differing only in terms of how much accompanying gestures they produced along with this talk.

These correlational data are consistent with the possibility that the gestures children see in the context of number talk promote their numerical understandings. However, in order to test this hypothesis, we need to carry out studies that experimentally manipulate children's exposure to gestures in the context of adults' verbal counts and their labeling of set size. This kind of approach is illustrated in Alibali and DiRusso's (1999) study of children's own points during counting and in several studies described in the next section that manipulate gestural input in the context of equivalence lessons (Church, Ayman-Nolley, & Mahootian, 2004; Ping & Goldin-Meadow, 2008; Valenzeno, Alibali, & Klatzky, 2003). Only one exploratory study has tried to use this approach to ask whether early number talk that is accompanied by gesture is more effective in promoting number development than early number talk in speech only. Suriyakham (2007) presented children whose cardinal number knowledge had been assessed on a pretest using the Give-a-Number task and the Point-to-X task with one of three types of training: (a) sets were counted, accompanied by a pointing gesture to each element, and then given a cardinal label; (b) sets were counted, accompanied by a pointing gesture to each element, and then given a cardinal label that was accompanied by a gesture that circled the entire set; and (c) the same training as (b) except that children also were asked to make a circling gesture to indicate the set size. Posttest assessments showed most growth for children in condition (a) possibly because the circling gesture, used to indicate that the cardinal label applied to the entire set rather than one object in the set, was an unfamiliar gesture, not found in naturalistic parent-child interactions; the gesture may not have been interpretable and perhaps was even distracting. The findings suggest that the juxtaposition of a pointing gesture in the context of counting and no hand movement (or at least no circling hand movement) in the context of labeling the cardinal value, a pattern that is seen frequently in naturalistic parent-child interactions, is particularly helpful to children in promoting understanding of cardinal number.

The Gestures That Teachers Produce During Mathematical Equivalence Lessons

Teachers routinely gesture in classrooms (Crowder & Newman, 1993; Flevares & Perry, 2001; Neill, 1991; Roth & Welzel, 2001; Zukow-Goldring, Romo, & Duncan, 1994), particularly if they are experienced (Neill & Caswell, 1993). This practice raises the question: Does gesture occur in math lessons, and, if so, is it used often enough to make a difference?

Goldin-Meadow, Kim, and Singer (1999) observed teachers conducting one-on-one individual tutorials with 9- and 10-year-old children who had not yet mastered mathematical equivalence with respect to addition. They found that the teachers expressed 40% of the problem-solving strategies they taught in gesture. When in a classroom situation, teachers also use gesture to convey their message. Flevares and Perry (2001) found that mathematics teachers used from five to seven nonspoken representations of mathematical ideas per minute (almost one every 10 seconds), and gesture was by far the most frequent nonspoken form for all of the teachers (the others were pictures, objects, and writing). Moreover, when the teachers combined two or more nonspoken representations, one of those forms was always a mathematically relevant gesture—gesture was the glue that linked the different forms of information to one another and to speech. Interestingly, the teachers often used their nonspoken representations strategically, responding to a student's confusion with a nonspoken representation. The teachers would repeat their own speech while clarifying the meaning of their utterance with gesture. This seemed to work—the children would then frequently come up with the correct answer.

However, these studies do not tell us whether the gestures that children see their teachers produce on a task lead children to improve their performance on that task the next time around. The few experimental studies that have been done suggest that children learn more from a lesson that contains gesture than from a lesson that does not contain gesture (Church, Ayman-Nolley, & Mahootian, 2004; Valenzeno, Alibali & Klatzky, 2003), even when the gestures are not directed at objects in the immediate environment (Ping & Goldin-Meadow, 2008). However, much more work needs to be done before we fully understand the conditions under which gesture promotes learning. Take, for example, the role of gesture-speech mismatches in instruction. We might have guessed that gesture would get in the way of learning when it conveys information that is different from the information conveyed in speech; that is, when it mismatches speech. We know that communication can suffer when speakers produce gestures that convey different information from their speech (Goldin-Meadow & Sandhofer, 1999; Kelly & Church, 1998; McNeil, Alibali, & Evans, 2000). Yet teachers frequently produce gesture-speech mismatches when teaching children who produce their own mismatches, and those children profit from their instruction (Goldin-Meadow & Singer, 2003). These findings suggest that mismatching gesture and speech could be good for learning. Indeed, in a study in which the instruction that children received in mathematical equivalence was experimentally manipulated, Singer and Goldin-Meadow (2005) found that presenting two strategies in a gesture-speech mismatch (one correct strategy in speech and a different, also correct strategy in gesture) was more effective than presenting those same two strategies entirely in speech. Perhaps in these instances gesture provides a representation that matches the child's next developmental state and, in this way, facilitates learning.

More generally, recommendations for mathematics curricula encourage teachers to present ideas through a variety of representations—diagrams, physical models, written text (National Council of Teachers of Mathematics, 1989). Shavelson, Webb, Stasz, and McArthur (1988), among others, recommend that teachers translate among alternative symbolic representations of a problem (e.g., math symbols and number line) rather than working within a single symbolic form. Gesture can serve as one of these representational formats, one that has a strong visual component. Gesture is unique, however, in that unlike a map or diagram, it is transitory—disappearing in the air just as quickly as speech. But gesture also has an advantage—it can be, indeed must be, integrated temporally with the speech it accompanies. We know that it is important for visual information to be timed appropriately with spoken information in order for it to be effective (Baggett, 1984; Mayer & Anderson, 1991). Thus, gesture used in conjunction with speech may present a more naturally unified picture to the student than a diagram used in conjunction with speech. If gesture were to become recognized as an integral—and inevitable—part of conversation in a teaching situation, it could perhaps be harnessed, offering teachers an excellent vehicle for presenting to their students a second perspective on the task at hand.

The Gestures Children Do

Counting

As noted earlier, many studies show that children tend to point to the objects in a set while counting them, and that this pointing behavior is correlated with more accurate counting. However, few studies have experimentally manipulated children's pointing gestures, which is necessary to test whether gesturing promotes better counting performance. In one study, 2-, 4,- and 6-year-olds were asked to count sets with two, three, seven, or eight objects. One group was encouraged to use pointing gestures while counting, and the other group was prevented from using gestures, which was achieved by displaying the objects in a cage that made it impossible for the participants to point while counting. Findings showed that the 4-year-olds' counting accuracy was significantly better in the gesture condition (Saxe & Kaplan, 1981). Alibali and DiRusso (1999) also explored gesture's role in children's counting by comparing three conditions: the child gestured while counting, the child was restricted from gesturing while counting, and the child watched a puppet gesture while counting. They found that children were most accurate in the two conditions in which counting was accompanied by gesture, theirs or the puppet's. But they were least likely to make errors coordinating number words and objects when they themselves produced the gestures. Thus, children's production of gestures while counting leads to earlier one-to-one correspondence between number words and objects than the use of number words alone. An important question is whether efforts to accelerate the accuracy of counting through the gesture accelerate the learning of associated mathematical concepts such as one-to-one correspondence and the cardinal principle.

Early Calculation

As we have seen, children's use of finger strategies when they are first learning to solve calculation problems is associated with performance accuracy. These correlational findings are consistent with the possibility that the gestures children produce when carrying out calculations scaffold their development, but we cannot conclude from these studies that gestures play a causal role in the development of calculation skill.

Researchers have suggested that children's finger use during calculation promotes their calculation skill. For example, Jordan et al. (2008) hypothesize that finger strategies help children progress from solving calculation problems with concrete objects to solving symbolic calculation problems. Siegler and Shipley (1995) made the more specific suggestion that finger counting when solving calculation problems can help children accurately make the association between number fact problems and their solutions. However, the critical experiments are yet to be done. For example, to determine whether finger counting in the context of calculation problems is causally related to children's later success in solving these problems through retrieval strategies, we need to experimentally manipulate children's finger strategies during calculation, as in Alibali and DiRusso's (1999) experiment on children's counting.

Mathematical Equivalence

The gestures that children see during a math lesson influence what they will learn from the lesson. It is likely that the information conveyed in a teacher's gestures has an impact on child learning, but the impact of teacher gesture on children may also be through child gesture—children who see their teachers gesture a concept are themselves likely to gesture the concept; in turn, these children are particularly likely to learn the concept (Cook & Goldin-Meadow, 2006). Indeed, evidence suggests that encouraging children to gesture prior to a mathematical equivalence lesson facilitates learning.

Broaders, Cook, Mitchell, and Goldin-Meadow (2007) asked children to explain how they solved six mathematical equivalence problems with no instructions about what to do with their hands. They then asked the children to solve a second set of comparable problems and divided the children into three groups: Some were told to move their hands as they explained their solutions to this second set of problems, some were told not to move their hands, and some were given no instructions about their hands. Children who were told to gesture on the second set of problems added strategies to their repertoires that they had not previously produced; children who were told not to gesture and children given no instructions at all did not. Most of the added strategies were produced in gesture and not in speech, and, surprisingly, most were correct. Telling children to gesture appeared to bring out ideas that they had not previously expressed. Moreover, articulating those ideas in gesture seemed to help the children become ready to learn about mathematical equivalence. When later given instruction in mathematical equivalence, children who were told to gesture, and added strategies to their repertoires, profited from the instruction and learned how to solve the math problems. Being told to gesture thus encouraged children to express ideas that they had previously not expressed, which in turn led to learning.

But can gesture, on its own, create new ideas? To determine whether gesture can create new ideas, we need to teach speakers to move their hands in particular ways. If these speakers can extract meaning from their own hand movements, they should be sensitive to the particular movements they are taught to produce and learn accordingly. Alternatively, all that may matter is that speakers move their hands. If so, they should learn regardless of which movements they produce. To investigate these alternatives, Goldin-Meadow, Cook, and Mitchell (2009) manipulated gesturing during a math lesson. They found that children required to produce correct gestures learned more than children required to produce partially correct gestures, who learned more than children required to produce no gestures. This effect was mediated by whether, after the lesson, the children added information to their spoken repertoire that they had conveyed only in their gestures during the lesson (and that the teacher had not conveyed at all). The findings confirm that gesture is involved not only in processing old ideas but also in creating new ones. This study suggests that we may be able to lay the foundations for new knowledge simply by telling learners how to move their hands (see Cook, Mitchell, & Goldin-Meadow, 2008, for related findings) or by moving our hands ourselves when teaching them (Cook & Goldin-Meadow, 2006).

Why might the act of gesturing facilitate learning? One possibility is that gesturing reduces demand on speakers' working memory, which in turn frees cognitive effort that can be applied to learning. If asked to remember an unrelated list of items while explaining how they solved a math problem, speakers are able to maintain more items in verbal working memory (and thus recall more items) when they gesture during the explanation than when they do not gesture. This effect has been found in both children and adults (Goldin-Meadow, Nusbaum, Kelly, & Wagner, 2001). The effect has also been found when speakers are asked to recall unrelated items in visual (rather than verbal) working memory (Wagner, Nusbaum & Goldin-Meadow, 2004); speakers maintain more items in visual working memory (and thus recall more items) when they gesture during their explanations than when they do not gesture, suggesting that gesturing does not lighten the load on working memory by transferring items from a verbal store to a visual store. In addition, gesturing reduces demand on working memory even when the gestures are not directed at visually present objects, that is, even when the objects are removed and the gestures are produced in space. If asked to remember an unrelated list of items while explaining how they solved conservation of quantity problems that elicit iconic gestures, speakers remember more items when they gesture during the explanation than when they do not gesture (Ping & Goldin-Meadow, 2010). Gesturing thus confers its benefits by more than simply tying abstract speech to objects directly visible in the environment.

Importantly, it is not just moving the hands that reduces demand on working memory—it is the fact that the moving hands convey meaning. Producing gestures that convey different information from speech (i.e., mismatches) reduces demand on working memory less than producing gestures that convey the same information in speakers who are experts on the task (Wagner et al., 2004). Somewhat paradoxically, we find the opposite effect in speakers who are novices—producing gestures that convey different information from speech (mismatches) reduces demand on working memory more than producing gestures that convey the same information as speech (Ping & Goldin-Meadow, 2010). This result could explain why mismatch is particularly good for learning in novices because a reduction in the demand on working memory would free up cognitive resources that can then be applied to the to-be-learned task. However, in both the expert and the novice, the meaning relation that the gesture holds to speech determines, at least in part, the extent to which the load on working memory is reduced. This saved effort can then be applied to grappling with new information.

THE ROLE OF THE BODY IN TEACHING ARITHMETIC

Hostetter and Alibali (2008) have proposed that gestures emerge from perceptual and motor simulations underlying the speaker's thoughts (see also Rimé & Schiaratura, 1991). This proposal is based on recent theories claiming that linguistic meaning is grounded in perceptual and action experiences (Barsalou, 1999; Glenberg & Kaschak, 2002; Richardson, Spivey, Barsalou, & McRae, 2003; Zwaan, Stanfield, & Yaxley, 2002). If so, gesture could be a natural outgrowth of the perceptual-motor experiences that underlie language. Under this view, the richer the simulations of action experiences, the more speech will be accompanied by gesture. For example, speakers tend to gesture more when describing dot patterns that they had constructed with wooden pieces than dot patterns that they had viewed on a computer screen (Hostetter & Alibali, 2010; see also Cook & Tanenhaus, 2009).

Focusing on acquisition and learning, Karmiloff-Smith (1992) described a process of redescription in which action representations are continuously redescribed into different representational formats. The process culminates in a verbal format that brings with it explicit awareness. Because gesture has its roots in action, it can exploit the effects that action has on thinking (cf. Beilock, Lyons, Mattarella-Micke, Nusbaum, & Small, 2008; see also Beilock & Goldin-Meadow, 2010; Goldin-Meadow & Beilock, 2010). But gesture is also a step removed from action and can thus selectively highlight components of action that are relevant to the to-be-learned problem and leave out components that are not relevant. As a result, gesture's contribution to learning may be used as a stepping-stone in the transition from concrete action to abstract thought.

One final point deserves mention with respect to action, gesture, and learning. Because gestures are external and can closely resemble the actions they represent (i.e., they can be relatively concrete), they have the potential to serve a scaffolding function comparable to the one that manipulatives serve in teaching math (Mix, 2010). Manipulatives are concrete objects designed to instantiate mathematical notions in external form and thus to off-load some of the mental effort involved in learning those notions (e.g., Cuisenaire rods—blocks that illustrate the decomposition of numbers, 1 + 6 = 2 + 5 = 3 + 4). They are typically used in the earliest stages of learning and are gradually replaced with more abstract symbols. Future work is needed to explore whether gesture's closeness to action does, in fact, contribute to its effect on learning and, if so, to determine how we can best manipulate gesture to enhance learning.

In summary, although mathematics is typically considered an abstract and a formal system, there is evidence of a dynamic component underlying this apparently static system in both learners and experts. Because of its close relation to action, gesture has the potential to capture this dynamic component and thus play a role in how arithmetic is taught and learned. The learner's own gestures may reveal an understanding of arithmetic not found in speech. Gesture can thus serve as a tool for recognizing when a child is ready to profit from instruction in a particular arithmetic notion. But gesture has the potential to do more than just reflect a child's knowledge of arithmetic. It can also play a role in changing that knowledge—the gestures that teachers produce can lead children to make inferences they would not have made if the teacher used speech alone, perhaps by bringing action (perceptual-motor) information into the mix. Gesture may thus serve as a pedagogical tool uniquely qualified to instill in children an understanding of arithmetic.

NOTE

1. Finger representations continue to play a role in numerical representations throughout the lifespan, even though adults rarely rely on finger counting to solve mathematical problems. Adults' finger counting habits (i.e., starting to count with the left hand versus the right hand) modulate the Spatial-Numerical Association of Response Codes (SNARC) effect—the association of spatial location and number, characterized by faster responses to lower numbers on the left and higher numbers on the right. In particular, the SNARC effect is reduced for those who start counting with the right hand rather than the left (Fischer & Zwaan, 2008). Supporting the role of finger counting in neural representations of number, findings from an event-related fMRI study showed that adults' motor areas contralateral to the hand used for counting small numbers are activated when small numbers or number words are presented, even when no overt hand movements are made (Hauk & Tshentscher, 2013).

REFERENCES

Alibali, M. W., & DiRusso, A. A. (1999). The function of gesture in learning to count: More than keeping track. *Cognitive Development, 14*(1), 37–56.

Alibali, M. W., & Goldin-Meadow, S. (1993). Gesture-speech mismatch and mechanisms of learning: What the hands reveal about a child's state of mind. *Cognitive Psychology, 25,* 468–523.

Baggett, P. (1984). Role of temporal overlap of visual and auditory material in forming dual media associations. *Journal of Educational Psychology, 76,* 408–417.

Barsalou, L. W. (1999). Perceptual symbol systems. *Behavioral and Brain Sciences, 22,* 577–660.

Beilock, S. L., & Goldin-Meadow, S. (2010). Gesture grounds thought in action. *Psychological Science, 21,* 1605–1610.

Beilock, S. L., Lyons, I. M., Mattarella-Micke, A., Nusbaum, H. C., & Small, S. L. (2008). Sports experience changes the neural processing of action language. *Proceedings of the National Academy of Sciences of the United States of America, 105,* 13269–13273.

Beller, S., & Bender, A. (2011). Explicating numerical information: When and how fingers support (or hinder) number comprehension and handling. *Frontiers of Psychology, 2,* Article 214.

Broaders, S. C., Cook, S. W., Mitchell, Z., & Goldin-Meadow, S. (2007). Making children gesture brings out implicit knowledge and leads to learning. *Journal of Experimental Psychology, 136*(4), 539–550.

Butterworth, G. (1999). Neonatal imitation: Existence, mechanisms and motives. In J. Nadel & G. Butterworth, (Eds.), *Cambridge studies in cognitive perceptual development* (pp. 63–88). New York, NY: University Press.

Church, R. B., Ayman-Nolley, S., & Mahootian, S. (2004). The role of gesture in bilingual instruction. *International Journal of Bilingual Education and Bilingualism, 7,* 303–319.

Church, R. B., & Goldin-Meadow, S. (1986). The mismatch between gesture and speech as an index of transitional knowledge. *Cognition, 23*(1), 43–71.

Cook, S., & Goldin-Meadow, S. (2006). The role of gesture in learning: Do children use their hands to change their minds? *Journal of Cognition and Development, 7,* 211–232.

Cook, S. W., Mitchell, Z., & Goldin-Meadow, S. (2008). Gesturing makes learning last. *Cognition, 106,* 1047–1058.

Cook, S. W., & Tanenhaus, M. K. (2009). Embodied communication: Speakers' gestures affect listeners' actions. *Cognition, 113,* 98–104.

Crollen, V., Seron, X., & Noël, M.-P. (2011). Is finger-counting necessary for the development of arithmetic abilities? *Frontiers in Psychology, 2,* Article 242.

Crowder, E. M., & Newman, D. (1993). Telling what they know: The role of gesture and language in children' s science explanations. *Pragmatics and Cognition, 1,* 341–376.

Fayol, M., Barrouillet, P., & Marinthe, C. (1998). Predicting arithmetical achievement from neuropsychological performance: A longitudinal study. *Cognition, 68,* B63–B70.

Fayol, M., & Seron, X. (2005). About numerical representations: Insights from neuropsychological, experimental, and developmental studies. In J. I. Campbell (Ed.), *Handbook of mathematical cognition* (pp. 3–22). New York, NY: Psychology Press.

Fischer, M. H., & Zwaan, R. A. (2008). *Grounding cognition in perception and action* [Special Issue of *Quarterly Journal of Experimental Psychology*]. Hove, England: Psychology Press.

Flevares, L. M., & Perry, M. (2001). How many do you see? The use of nonspoken representations in first-grade mathematics lessons. *Journal of Educational Psychology, 93,* 330–345.

Fuson, K. C. (1988). *Children's counting and concepts of number.* New York, NY: Springer-Verlag.

Fuson, K. C., & Hall, J. W. (1983). The acquisition of early number word meanings: A conceptual analysis and review. In H. P. Ginsburg (Ed.), *The development of mathematical thinking* (pp. 49–107). New York, NY: Academic Press.

Gelman, R., & Gallistel, C. R. (1978). *The child's understanding of number.* Cambridge, MA: Harvard University Press.

Gelman, R., & Meck, E. (1983). Preschoolers' counting: Principles before skill. *Cognition, 13,* 343–359.

Glenberg, A. M., & Kaschak, M. P. (2002). Grounding language in action. *Psychonomic Bulletin & Review, 9,* 558–565.

Goldin-Meadow, S. (2003). *Hearing gesture: How our hands help us think.* Cambridge, MA: Harvard University Press.

Goldin-Meadow, S., & Beilock, S. L. (2010). Action's influence on thought: The case of gesture. *Perspectives on Psychological Science, 5,* 664–674.

Goldin-Meadow, S., Cook, S. W., & Mitchell, Z. A. (2009). Gesturing gives children new ideas about math. *Psychological Science, 20*(3), 267–272.

Goldin-Meadow, S., Kim, S., & Singer, M. (1999). What the teacher's hands tell the student's mind about math. *Journal of Educational Psychology, 91*(4), 720–730.

Goldin-Meadow, S., Nusbaum, H., Kelly, S. D., & Wagner, S. (2001). Explaining math: Gesturing lightens the load. *Psychological Science, 12*(6), 516–522.

Goldin-Meadow, S., & Sandhofer, C. (1999). Gestures convey substantive information about a child's thoughts to ordinary listeners. *Developmental Science, 2,* 67–74.

Goldin-Meadow, S., & Singer, M. (2003). From children's hands to adults' ears: Gesture's role in the learning process. *Developmental Psychology, 39,* 509–520.

Gracia-Bafalluy, M., & Noël, M.-P. (2008) Does finger training increase young children's numerical performance? *Cortex, 44*(4), 368–375.

Graham, T. A. (1999). The role of gesture in children's learning to count. *Journal of Experimental Child Psychology, 74,* 333–355.

Hauk, O., & Tschentscher, N. (2013). The body of evidence: What can neuroscience tell us about embodied semantics? *Frontiers in Psychology, 4,* Article 50.

Hostetter, A. B., & Alibali, M. W. (2008). Visible embodiment: Gestures as simulated action. *Psychonomic Bulletin & Review, 15,* 495–514.

Hostetter, A. B., & Alibali, M. W. (2010). Language, gesture, action! A test of the gesture as simulated action framework. *Journal of Memory and Language, 63,* 245–257.

Jordan, N. C., Huttenlocher, J., & Levine, S. C. (1992). Differential calculation abilities in young children from middle- and low-income families. *Developmental Psychology, 28,* 644–653.

Jordan, N. C., Kaplan, D., Ramineni, C., & Locuniak, M. N. (2008). Development of number combination skill in the early school years: When do fingers help? *Developmental Science, 11*(5), 662–668.

Jordan, N. C., Levine, S. C., & Huttenlocher, J. (1994). Development of calculation abilities in middle- and low-income children after formal instruction in school. *Journal of Applied Developmental Psychology, 15,* 223–240.

Karmiloff-Smith, A. (1992). *Beyond modularity: A developmental perspective on cognitive science.* Cambridge, MA: MIT Press.

Kelly, S. D., & Church, R. B. (1998). A comparison between children's and adults' ability to detect conceptual information conveyed through representational gesture. *Child Development, 69*(1), 85–93.

Lakoff, G., & Nunéz, R. (2000). *Where mathematics comes from: How the embodied mind brings mathematics into being.* New York, NY: Basic Books.

Le Corre, M., Van de Walle, G., Brannon, E. M., & Carey, S. (2006). Re-visiting the competence/performance debate in the acquisition of the counting principles. *Cognitive Psychology, 42,* 130–169.

Levine, S. C., Jordan, N. C., & Huttenlocher, J. (1992). Development of calculation abilities in young children. *Journal of Experimental Child Psychology, 53,* 72–93.

Levine, S. C., Suriyakham, L., Rowe, M., & Huttenlocher, J., & Gunderson, E. A. (2011). What counts in the development of children's number knowledge? *Developmental Psychology, 46,* 1309–1313.

Marghetis, T., & Núñez, R. (2010). *Dynamic construals, static formalisms: Evidence from co-speech gesture during mathematical proving* (Center for Research on Language Technical Report, Vol. 22, No. 1). San Diego, CA: University of California at San Diego.

Marinthe, C., Fayol, M., & Barrouillet, P. (2001). Gnosies digitales et développement des performances arithmétiques. In A. Van Hout, C. Meljac, & J. P. Fischer (Eds.), *Troubles du calcul et dyscalculies chez l'enfant* (pp. 239–254). Paris, France: Masson.

Mayer, R. E., & Anderson, R. B. (1991). Animations need narrations: An experimental test of a dual-coding hypothesis. *Journal of Educational Psychology, 83,* 484–490.

McNeil, N., Alibali, M. W., & Evans, J. L. (2000). The role of gesture in children's comprehension of spoken language: Now they need it, now they don't. *Journal of Nonverbal Behavior, 24,* 131–150.

McNeil, N. M., Grandau, L., Knuth, E. J., Alibali, M. W., Stephens, A. S., Hattikudur, S., & Krill, D. E. (2006). Middle-school students' understanding of the equal sign: The books they read can't help. *Cognition and Instruction, 24,* 367–385.

Mix, K. S. (2010). Spatial tools for mathematical thought. In S. Mix, L. B. Smith, & M. Gasser (Eds.), *The spatial foundations of cognition and language: Thinking through space* (pp. 40–66). Oxford, England: Oxford University Press.

National Council of Teachers of Mathematics. (1989). *Curriculum and evaluation standards for school mathematics.* Reston, VA: Author.

Neill, S. (1991). *Classroom nonverbal communication.* London, England: Routledge.

Neill, S., & Caswell, C. (1993). *Body language for competent teachers.* London, England: Routledge.

Nicoladis, E., Pika, S., & Marentette, P. (2010). Are number gestures easier than number words for preschoolers. *Cognitive Development, 25,* 247–261.

Noël, M.-P. (2005). Finger gnosia: A predictor of numerical abilities in children? *Child Neuropsychology, 11*(5), 413–430.

Núñez, R. (2008). Mathematics, the ultimate challenge to embodiment: Truth and the grounding of axiomatic systems. In O. Calvo & T. Gomila (Eds.), *Handbook of cognitive science* (pp. 333–353). Amsterdam, the Netherlands: Elsevier Science.

Perry, M., Church, R. B., & Goldin-Meadow, S. (1988). Transitional knowledge in the acquisition of concepts. *Cognitive Development, 3,* 359–400.

Ping, R., & Goldin-Meadow, S. (2010). Gesturing saves cognitive resources when talking about nonpresent objects. *Cognitive Science, 34,* 602–619.

Ping, R. M., & Goldin-Meadow, S. (2008). Hands in the air: Using ungrounded iconic gestures to teach children conservation of quantity. *Developmental Psychology, 44,* 1277–1287.

Potter, M. C., & Levy, E. I. (1968). Spatial enumeration without counting. *Child Development, 39,* 265–272.

Richardson, D. C., Spivey, M. J., Barsalou, L. W., & McRae, K. (2003). Spatial representations activated during real-time comprehension of verbs. *Cognitive Science, 27*, 767–780.

Rimé, B., & Schiaratura, L. (1991). Gesture and speech. In R. S. Feldman & B. Rimé (Eds.), *Fundamentals of nonverbal behavior* (pp. 239–281). New York, NY: Cambridge University Press.

Roth, W.-M., & Welzel, M. (2001). From activity to gestures and scientific language. *Journal of Research in Science Teaching, 38*, 103–136.

Saxe, G. B. (1977). A developmental analysis of notational counting. *Child Development, 48*, 1512–1520.

Saxe, G. B., & Kaplan, R. (1981). Gesture in early counting: A developmental analysis. *Perceptual & Motor Skills, 53*, 851–854.

Schaeffer, B., Eggleston, V., & Scott, J. (1974). Number development in young children. *Cognitive Psychology, 6*, 353–379.

Shavelson, R. J., Webb, N. M., Stasz, C., & McArthur, D. (1988). Teaching mathematical problem solving: Insights from teachers and tutors. In R. I. Charles & E. A. Silver (Eds.), *The teaching and assessing of mathematical problem solving* (pp. 203–231). Reston, VA: National Council of Teachers of Mathematics.

Siegler, R. S., & Jenkins E. (1989). *How children discover new strategies.* Hillsdale, NJ: Lawrence Erlbaum Associates.

Siegler, R. S., & Shipley, C. (1995). Variation, selection, and cognitive change. In T. Simon & G. Halford (Eds.), *Developing cognitive competence: New approaches to process modeling* (pp. 31–76). Hillsdale, NJ: Lawrence Erlbaum Associates.

Singer, M. A., & Goldin-Meadow, S. (2005). Children learn when their teacher's gestures and speech differ. *Psychological Science, 16*(2), 85–89.

Suriyakham, L. W. (2007). Input effects on the development of the cardinality principle: Does gesture count? (UMI No. 3262306). Retrieved April 21, 2011, from Dissertations and Theses database.

Valenzeno, L., Alibali, M. W., & Klatzky, R. (2003). Teachers' gestures facilitate students' learning: A lesson in symmetry. *Contemporary Educational Psychology, 28*, 187–204.

Wagner, S. M., Nusbaum, H., & Goldin-Meadow, S. (2004). Probing the mental representation of gesture: Is handwaving spatial? *Journal of Memory and Language, 50*, 395–407.

Wiese, H. (2003). Iconic and non-iconic stages in number development. *Trends in Cognitive Sciences, 7*(9), 385–390.

Wynn, K. (1990). Children's understanding of counting. *Cognition, 36*, 155–193.

Zukow-Goldring, P., Romo, L., & Duncan, K. R. (1994). Gestures speak louder than words: Achieving consensus in Latino classrooms. In A. Alvarez & P. del Rio (Eds.), *Education as cultural construction: Exploration in socio-cultural studies* (Vol. 4, pp. 227–239). Madrid, Spain: Fundacio Infancia y Aprendizage.

Zwaan, R. A., Stanfield, R. A., & Yaxley, R. H. (2002). Language comprehenders mentally represent the shape of objects. *Psychological Science, 13*, 168–171.

SECTION II

GESTURE AND EMBODIMENT IN SCHOOL MATHEMATICS

CHAPTER 4

ANALYTIC-STRUCTURAL FUNCTIONS OF GESTURES IN MATHEMATICAL ARGUMENTATION PROCESSES

Ferdinando Arzarello
Cristina Sabena
Università di Torino, Italy

In our research on gestures in mathematical understanding, we start by assuming that gestures, along with other body aspects (e.g., gazes), are an important component of emotional and cognitive manifestations and an inseparable component of speech discourse, as research in psychological and related domains have shown (e.g., Goldin-Meadow, 2003; McNeill, 1992, 2005). From our own observation in the classrooms and from recent literature, we have become increasingly convinced that gestures play important roles in the mathematics classroom when: students solve problems (Radford, 2010), students and teachers interact (Arzarello, Paola, Robutti, & Sabena, 2009), students explain a mathematical concept (Edwards, 2009), or the teacher lectures (Pozzer-Ardenghi & Roth, 2008). In all such cases, gestures are considered not as isolated variables but rather as part of the

Emerging Perspectives on Gesture and Embodiment in Mathematics, pages 75–103

multisemiotic arsenal at the students' disposal to bridge the gap between everyday experience and formal mathematics.

Keeping these premises as our starting points, in this chapter, we focus on the gesture-speech relationship and study their contribution in structuring a mathematical argumentation. Our goal is to show how gestures can contribute not only to the semantic content of mathematical ideas but also to the logical structure that organizes them in mathematical arguments. We will base our analysis on the results and methodologies from three domains, which are presented in the next section:

- gesture studies, mainly referring to McNeill's frame on the gesture-language relationship and the notion of catchment;
- the analytic-linguistic frame given by Chomsky; and
- the prosodic element of intonation of a phrase.

We then substantiate our claim through an episode from a case study related to a fifth-grade problem-solving activity involving 10-year-old Italian students. The episode will be analyzed in detail by integrating typical methodologies of gesture studies with the analytic-linguistic and prosodic ones. A final discussion will reflect on the implications of this research.

THEORETICAL FRAMEWORK

The Language-Gesture Relationship

In his seminal work, McNeill (1992) defines gestures as "the movements of the hands and arms that we see when people talk" (McNeill, 1992, p. 1). This approach comes from the analysis of conversational settings and has been widely adopted in successive research studies in psychology, in which gestures are viewed as distinct but inherently linked with speech utterances. Indeed, gestures and speech are considered as complementary sides of the same coin, and the contrast between the syntactic and morphologic aspects of speech versus the holistic properties of gesticulation is often highlighted. Linguistic systems are described as conveying meaning in a segmented, analytic, linear, and hierarchically structured way; gestures, on the contrary, are characterized as global, synthetic, multidimensional, and never hierarchical (McNeill, 1992). In his semiology, Saussure (1916) attributes the linear-segmented character of spoken language to its unidimensionality, contrasted to the multidimensionality of meanings. This characterization is taken up by Goldin-Meadow (2003), who contrasts language and gesture features:

> Language can only vary along the single dimension of time. [...] This restriction forces language to break meaning complexes into segments and to

> reconstruct multidimensional meanings by combining the segments in time. But gesture is not similarly restricted. Gestures are free to vary on dimensions of space, time, form, trajectory, and so on, and can present meaning complexes without undergoing segmentation or linearization. (pp. 24–25)

Language in fact is made of elementary constituents, and its rules of production involve both atomic (single) and molecular (compound) signs. On the contrary, gestures are described as endowed with holistic features that cannot be split into atomic components. The modes of production and transformation are often idiosyncratic[1] to the subject who produces them. Furthermore, one gesture can combine many meanings (synthetic character), and two gestures produced together do not combine to form a larger, more complex, hierarchically higher gesture, as language, in contrast, does (McNeill, 1992).

Besides specifying their inherent differences, research in psychology informs us about an intimate unity between speech and gesture. Empirical evidence shows in fact that they are semantically and pragmatically co-expressive. They are essentially synchronous in time and meaning, and they develop together in children. In quite expressive terms, McNeill (1992) claims that gestures have an active constitutive role on thought: "Gestures do not just reflect thought but have an impact on thought. Gestures, together with language, *help constitute thought*" (p. 245; italics original).

In this chapter, we consider the mathematical activity of argumentation and study the contribution of language and gestures in constituting not only the semantic content but also the structure of the argumentation. For our purposes, we will base our work on McNeill's framework on gestures and in particular on his notion of catchment. A catchment is recognizable when some gesture form features are seen to recur in at least two (not necessarily consecutive) gestures. According to McNeill, a catchment indicates discourse cohesion, and it is due to the recurrence of consistent visuospatial imagery in the speaker's thinking. Catchments may, therefore, be of great importance, giving us information about the underlying meanings in a discourse and about their dynamics.

> By discovering the catchments created by a given speaker, we can see what this speaker is combining into larger discourse units—what meanings are being regarded as similar or related and grouped together, and what meanings are being put into different catchments or are being isolated, and thus are seen by the speaker as having distinct or less related meanings. (McNeill et al., 2001, p. 10)

Furthermore, catchments may also give clues about the organization of the discourse at a meta-narrative level (McNeill, 2005). We will develop further this characteristic of gestures, which appears linked to the logical aspects of discourse, in presenting our case of mathematical discourse.

Elements of Linguistic Analysis

In the previous section, we have seen some major features of gestures that distinguish them from language; we can summarize these features with the words of Sandler (2009):

1. "[T]he gestures [...] are holistic, noncombinatoric, idiosyncratic, and context-sensitive. The linguistic signal is the converse: dually patterned, combinatoric, conventionalized, and far less context-dependent in the relevant sense" (p. 268).
2. Gestures and speech combine "in a simultaneous and complementary fashion linguistically organized material with holistic and iconic expressions. The two modalities participate in what we may call symbolic symbiosis" (p. 242).

These specific modalities of language are accomplished through its syntactical rules: recursion,[2] combinatoricity, conventionality, and so on (Hauser, Chomsky, & Fitch, 2002; Sandler, 2009), that is, all those properties that we call here analytic-structural.

In this section, we will introduce through examples some of the analytic-structural properties of language, which differentiate it from gestures, at least at a first analysis. A typical example of the recursive nature of language is when a phrase is nested within another phrase of the same type, as in the sentence, "I told John the information that Mary told me." The word *recursive* indicates two main intertwined aspects of language, which differentiate it from gestures (see Bara & Tirassa, 1999). First, language is compositional; that is, it is made up from atomic constituents through a recursive process that generates its more complex components from phonemes to words to sentences to discourses in natural languages. Also many specific mathematical languages have the same features. For example in elementary algebra, a complex expression is built up from atomic constituents—for example, $2x + 3(x^2 - 5)$ is built from x, the numbers, and the arithmetic operators. Second, this recursive process generates complex components that have a tree-like structure as displayed in Figure 4.1.

The tree representation makes evident the analytic-structural aspects of a sentence or a mathematical formula. When considering this facet, all contextual and imaginative traits are put in brackets. Only the logical and syntactic structure is considered. In some cases, an analytic-structural enquiry is sufficient for establishing the truth of a proposition. For example, the sentences

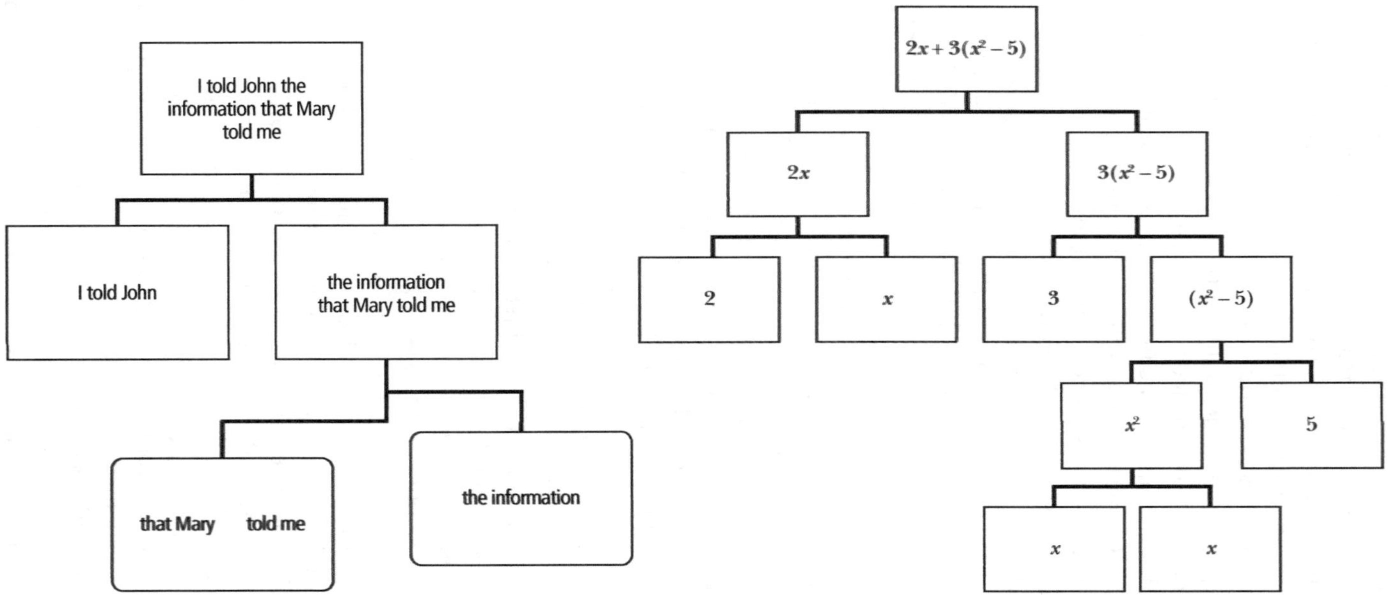

Figure 4.1 The tree-analysis of a sentence (“I told John the information that Mary told me”) and of a mathematical formula (“$2x + 3[x^2 - 5]$”).

"he will arrive or will not arrive in 50 days" and "all bachelors are unmarried" are true because of purely analytic reasons, namely, the meanings of "or" and "bachelor" in English. The reader may contrast these examples with the sentence "some bachelors are sad," where the purely analytic-structural features are not enough to establish the truth-falsity of the sentence.

A further ingredient of structural linguistics, which we use in our analysis, is the distinction between deep and surface structure of a linguistic production, which is a major principle in the grammatical theory of Noam Chomsky (1965). Chomsky's idea is that each sentence in a language has two levels of representation—a deep structure and a surface structure. In a nutshell, the deep structure is a direct representation of the basic semantic relations underlying a sentence and is mapped onto the surface structure, which follows the phonological form of the sentence closely via transformations. For example, the sentences "John loves Mary" and "Mary is loved by John" show differences at the surface structure, but they share the same deep structure.

Last but not least, in the list of our framework ingredients is the so-called prosodic analysis. Prosody concerns the stress, rhythm, and intonation of speech. We will now briefly illustrate them. Each word has a main stress. For example, in "beautiful" or "angolo" (angle in Italian), the stress is on the first syllable, which is followed by two unstressed syllables. Words in a sentence are combined according to particular rhythms. For example, in "good bye," the rhythm is different with respect to "not now" (the former is a "iamb," where only the second word is typically accented; the latter is a "spondee," where both words are usually accented). The rhythm of a sentence is particularly evident in, but not limited to, poetry.[3] In a sentence, we can also have different intonations (Hirst & Di Cristo, 1998). For example, the interrogative sentence "beve caffé?" ("Does she drink coffee?") has a different intonation than the corresponding declarative sentence "beve caffé" ("She drinks coffee"). In spoken Italian, there is no such word as "do" in English to indicate the interrogative form.

In our analysis, we will limit ourselves to consider only intonations, simply distinguishing between rising and falling intonations, that is, when the pitch of the voice increases or decreases, respectively, over time. For example, you can give two different intonations to the sentence "Why don't you move to Rome?" You could say it to ask a question (falling intonation in "don't") or to give a suggestion (rising intonation in "move"). We will also point out so-called dipping points (when a rising intonation follows a falling one) and peaking points (the opposite).

Taking McNeill and his collaborators' (2001) stance that the "organization of discourse is inseparable from gesture and prosody" (p. 30), we will integrate the analytic-linguistic, gesture, and intonation analysis to show that:

- when communicating mathematical ideas, gestures also can incorporate the analytic-structural functions typical of speech language; and
- considering the prosody of speech, we may get a deeper insight into the way the unity speech-gestures provide a structure to an argumentative discourse.

In other words, in mathematical argumentation processes, the two components, gestures and speech, may show a strong unity not only at semantic level but also at a structural one. In the next section, we illustrate our claims by analyzing in some detail a paradigmatic example from a case study in primary school.

INTEGRATING GESTURE, STRUCTURAL, AND PROSODIC ANALYSIS

The Case Study

We refer to a case study involving a group of fifth-grade children who were solving a problem inspired by the legend of Penelope from Homer's *Odyssey* (Arzarello et al., 2006; Bazzini, Sabena & Villa, 2009) presented through this story:

> On the island of Ithaca, Penelope had been waiting ten years for the return of her husband Ulysses from the war. On Ithaca, however, a lot of men wanted to take the place of Ulysses and marry Penelope. One day the goddess Minerva told Penelope that Ulysses was returning and his ship would arrive at Ithaca in 50 days. Penelope immediately summoned the suitors and told them:
>
> "I have decided, I will choose from you my bridegroom and the wedding will be celebrated when I have finished weaving a new piece of cloth for the nuptial bed. I will begin today and I promise to weave every two days; and when I have finished, the cloth will be my dowry." The suitors accepted. The cloth had to be 15 spans in length. Penelope immediately began to work, but one day she wove a span of cloth, while the following day, in secret, she undid half a span....
>
> Will Penelope choose another bridegroom? Why?

The problem-solving phase was the first step in a teaching experiment aimed at introducing pupils to modeling situations involving changing variables in arithmetic contexts and to related mathematical representations (such as tables and graphs). It is beyond the scope of this chapter to describe and discuss the students' solution strategies. However, we want to stress that the problem solution is not intuitive; pupils need to figure out the situation, make comparisons among data, and make computations to reach the final result. Penelope, weaving a half span of cloth every two days, after 50 days will have produced only 12.5 spans, so she will not have to marry another bridegroom because Ulysses will have already arrived. Students were allowed to approach the solution in any manner. Drawing was extensively used as a means of expression. Many students used concrete materials such as pens (a collection of 50 pens representing the 50 days) and paper (representing the Penelope's cloth). However, the key passages to the solution were strongly supported by the use of discourse accompanied by gestures and written representations (Bazzini, Sabena, & Villa, 2009).

Our analysis will focus on the last phase of the solving process of a group of five children who will be referred to by the pseudonyms Davide, Edoardo, Maria, Ornella, and Simona. The teacher did not intervene during the group work; the children were familiar with the work, and they knew that at the end they had to write down a unique, shared solution (which later was meant to be discussed with the other classroom groups). The five students worked together for about 20 minutes and reached a shared solution, first by finding the "local rule" of "four days for a span" and then organizing the information on a table to get the global result. We focus on the final phase, in which, after solving the problem, some children have difficulties stating the answer of the story. We relate how a girl, Simona, was able to provide a conclusion and correctly argue it by means of a gesture-speech synergy in which gestures had a fundamental role in expressing analytic-structural aspects.

For the sake of clarity, we present our analysis in three steps. The first step focuses on the student's uttered speech and makes use of the linguistic analysis tools. The second step analyses the gestures accompanying Simona's argument. Finally, the third step integrates the previous analysis with the prosodic component. The three steps consider separately three components of the unit speech gesture only for the sake of the analysis; however, the three components constitute a compact core of that unit.

Analytic-Structural Analysis

Simona's argument is contained in line 13 of the following excerpt. For situating it within the ongoing group work, we report the full excerpt in which the solution of the problem is finally stated by the children. The excerpt lasted 69 seconds and is presented here in English translated from the original Italian.

1. **Maria:** When Ulysses... if she ever takes away... eh... takes away some spans and however there are always... when Ulysses has arrived there are still... so...
2. **Maria & Simona** [*together overlapping*]: So the cloth is not yet finished...
3. **Maria:** And she could continue it...
4. **Simona:** So...
5. **Davide:** Exactly! But Ulysses had already arrived, so...
6. **Ornella:** So she could not finish the cloth...
7. **Davide:** 'Cause in 64 days she had done only 12...
8. **Simona:** Wait! She did not marry the same however... [*not understandable and overlapped with Ornella*]
9. **Ornella:** Because she was able... she had not been able to do the cloth...
10. **Edoardo:** but she... so
11. **Simona:** So she did not marry anyone.
12. **Davide:** Since she did not finish the cloth, she did not marry any of the suitors.
13. **Simona:** If Ulysses came before... if Ulysses came before 50 days... and she had already completed the cloth, she married another one. And on the contrary if... she... she did not finish... she did not finish making the cloth within 50 days, she took back Ulysses, who was already arrived... hence the suitorssss... could not marry her anymore.

We see that all the students try to give a global sense to the results they got with their previous trials, conjectures, and computations. Each one makes a partial contribution, although not always a correct one (like Davide in line 7). At the end (lines 11 and 13), Simona draws the conclusion by formulating the solution of the problem in a logical, well-organized synergy of words and gestures. It is important to observe that this formulation is produced after the elaboration of the solution. In such a process, crucial roles were played by the narrative dimension, sustained by both speech and gestures.[4]

In her final argument (lines 11 and 13), Simona goes beyond the narrative dimension and gives reasons for all the logical possibilities in the story. She explains why the story plot makes sense; her argument here is a sort of logical certification grounded on all the facts that happened in the story and on the group's investigations of it. It represents a step further toward the logical dimension of the story. Previously, the students addressed what happened and how it happened; now Simona explains why it could happen in the way it did.

Drawing inspiration from Chomsky's frame, we emphasize a specific kind of deep structure of particular interest in sentences used in mathematical arguments: their deep logical structure. For example, the two sentences "if a triangle has two congruent sides, it has also two congruent angles" and "in an isosceles triangle, two angles are congruent" have the same deep logical structure. Let us analyze the logical structure of Simona's argument.

In her discourse, Simona first states the claim—line 11: "Hence she did not marry anyone"—and then properly sustains it with a logical argument by cases. To dissect the argument, we use an analytic-structural lens, which points out its deep structural-logical features. Hence, we rewrite it divided according to its logical constituents. Numbers in parenthesis indicate the elapsed time in seconds when each utterance occurred.

13. **Simona:**

  a. (43") If Ulysses came before...

  b. (47") if Ulysses came before 50 days...

  c. (49") and she had already completed the cloth

  d. (51") she married another one.

  e. (53") And on the contrary if...

  f. (55") she... she did not finish... she did not finish making the cloth within 50 days

  g. (60") she took back Ulysses,

  h. (62") who was already arrived (64")...

  i. (66") hence the suitorssss... could not marry her anymore (69").

In her argument, Simona identifies a crucial condition, according to which two cases can be distinguished: whether, at Ulysses's arrival, Penelope has or has not finished the cloth. Hence, utterances in lines 13a–d contain the first case, and lines 13f–i contain the second case.

Simona starts by considering the most dramatic case: if Ulysses arrives (before 50 days) and Penelope has already completed her work (13a, b, c). In this case, Penelope has to marry one of the suitors (13d). Then, Simona considers the other case—that is, the case in which Ulysses arrives (before 50 days), and she has not yet completed the cloth (13f, h). So, she can "take back" her husband, and the suitors are out of the running (13g, i). Coding the argument in logical terms (and referring to line a–i segmentation of line 13):

[if (b & c) then d] & [if (f & h) then (g & i)]

This is logically equivalent to:

[if (b & c) then d] & [if [b & (not c)] then [g & (not d)]]

This is a mathematically perfect argument by cases. In fact, from a logical point of view, (i) is equivalent to the negation of (d): the equivalence is a transformation from a sentence where Penelope marries a suitor to one where no suitor marries Penelope. From the other side, (d) implies also not (g) because of marriage laws. Figure 4.2 shows the tree structure of the argument.

We observe also that in Simona's sentence, logical features are intermingled with narrative features; the latter add a semantic connotation to the deep logical structure of her argument. In fact, two opposite situations are presented as two possible endings of the story, a narrative technique sometimes used in cinematographic movies. After a first scene showing Ulysses's arrival, the final scenes would be either Penelope forced to marry a man she does not want or the suitors taking flight (see for instance the different subjects of the sentences). The two sentences "Penelope could not marry the suitors anymore" and "The suitors could not marry her anymore" are logically equivalent (i.e., they have the same deep structure), but the first one does not make sense within the story. In other words, the narrative plane provides the sense for the whole logical argumentation. As pointed out by Donaldson (1978/2006) in her critical discussion about Piaget's classical "three mountain problem," even very young children are able to solve a challenging problem, provided it has a "human sense"; that is, it allows children to act and imagine according to human fundamental goals and interactions (in Penelope's case, escaping from a second, undesired marriage).

Of course the narrative plane, which was in the foreground when the solution had to be produced, is now in the background. In fact, focused as they are on the logical relationships, the children can even partially modify the content of the story, without realizing it.[5]

Gesture Analysis

We will turn now to consider the gestures that Simona performs in her argument and illustrate their contribution in coordinating its semantic and logical aspects. We report the gestures in Table 4.1, trying to illustrate their dynamics by means of sequences of pictures and short descriptions. For the reader's sake, we indicate time information (elapsed time, in seconds, from the beginning of the episode), the co-timed utterances (included the original ones in Italian, in which underlined words are co-timed with shown gestures), and a code that will be used in the subsequent analysis.

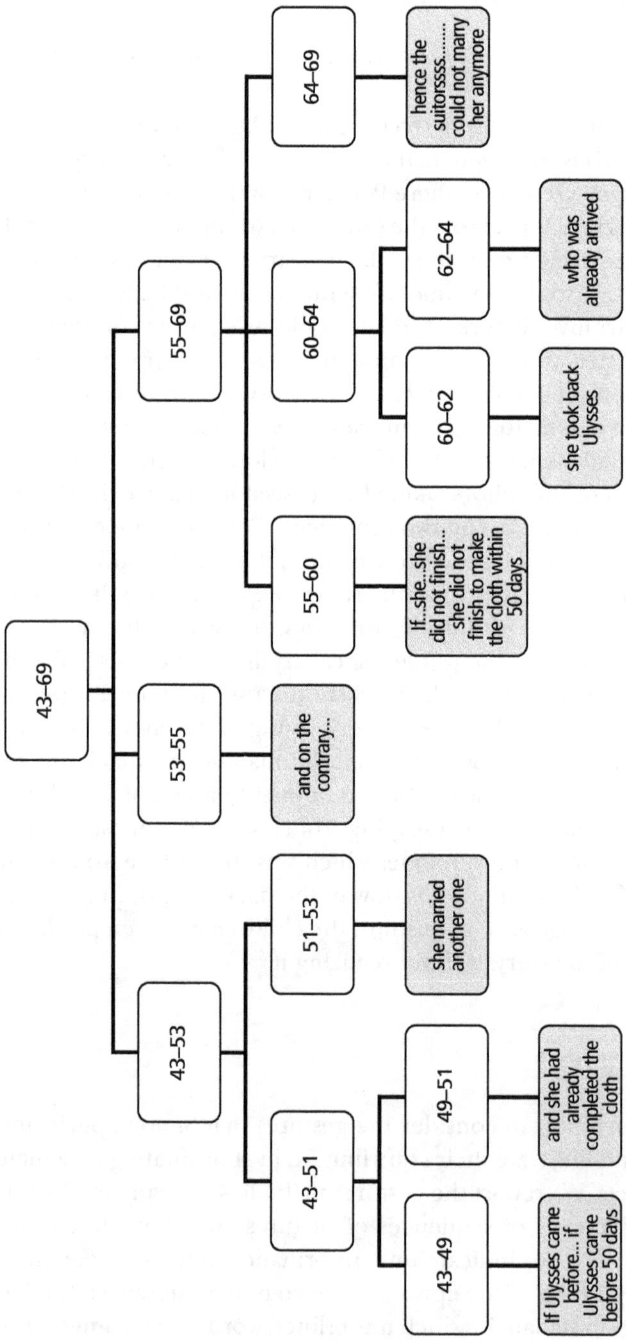

Figure 4.2 The tree structure of sentences in 13 obtained with a logical-linguistic analysis (numbers refer to time in seconds).

TABLE 4.1 Simona's Argument, Carried out by a Synergy of Speech and Gestures

Speech	Gesture	Code
(43") **a.** If Ulysses came before . . . [Se venìva <u>Ulìsse</u> prìma]	 **Figure 4.3** Pointing gestures to the written text of the story.	↓
(47") **b.** if Ulysses came before 50 days . . . [se venìva Ulìsse èntro <u>cìnquànta</u> giòrni . . .]	 **Figure 4.4 a, b, c** Beats rightward.	| *r*

(continued)

TABLE 4.1 Simona's Argument, Carried out by a Synergy of Speech and Gestures (continued)

Speech	Gesture	Code
(49") **c.** and she [<u>e lèi</u>]	 **Figure 4.5 a, b** Left hand suspended in the center	—*l*
had already completed the cloth [<u>avèva già finìto di fàre la copèrta</u>]	 **Figure 4.6 a, b, c** Right hand beating.	⎝*r*

(continued)

TABLE 4.1 Simona's Argument, Carried out by a Synergy of Speech and Gestures (continued)

Speech	Gesture	Code
	 **Figure 4.6 d, e** Right hand beating.	|r
(51") **d.** she married another one [lèi sposàva un àltro]	 **Figure 4.7 a, b, c** Beats and pointings toward the left part of the gesture space, with the right hand.	⇐

(continued)

TABLE 4.1 Simona's Argument, Carried out by a Synergy of Speech and Gestures (continued)

Speech	Gesture	Code
(53") **e.** And on the contrary if… [e invèce se]	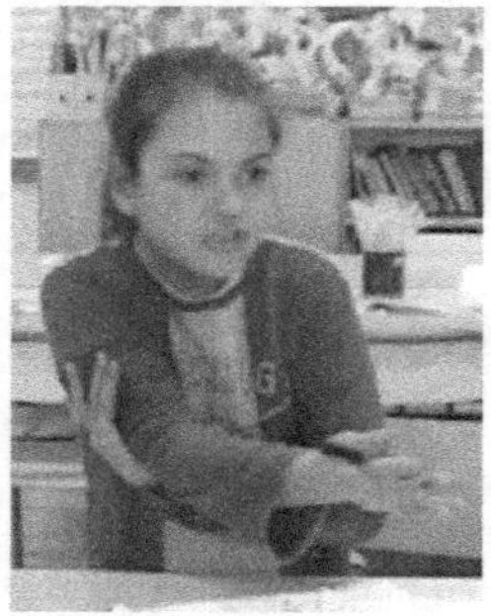 **Figure 4.8** Beat in the central part of the gesture space: the right arm held up by the left one.	⇑
(55") **f.** she…she did not finish…she did not finish making the cloth within 50 days [lèi…lèi nòn finìva…lèi nòn finìva di fàre la copèrta èntro cinquànta giòrni]	 **Figure 4.9 a, b, c, d** Beats performed moving both the hands first in the center and then to the right part of the gesture space.	⋃ ⇒

(continued)

TABLE 4.1 Simona's Argument, Carried out by a Synergy of Speech and Gestures (continued)

Speech	Gesture	Code
(60") **g.** she took back Ulysses, [lèi si prendèva di nuòvo <u>Ulìsse</u>]	 **Figure 4.10** Pointing to the sheet containing the text of the story (repeated four times)	↓
(62") **h.** who was already arrived (64") . . . [<u>che èra già arrivàto</u>]	 **Figure 4.11 a, b** Beats with both the hands, slightly rightward.	⋃

(continued)

TABLE 4.1 Simona's Argument, Carried out by a Synergy of Speech and Gestures (continued)

Speech	Gesture	Code
(66") **i.** hence the suitorssss . . . [quìndi i pretendèntiiii . . .]	 **Figure 4.12** The left hand points to the text.	↓
. . . could not marry her anymore **(69").** [nòn potevàno più sposàrla]	 **Figure 4.13** The left hand rises upward, waving in the air.	↗

(continued)

TABLE 4.1 Simona's Argument, Carried out by a Synergy of Speech and Gestures (continued)

Speech	Gesture	Code

Notes: Underlined text indicates a co-timed gesture or gestures.

Symbols in the third column indicate:

↓	beat downward
)l	suspended open left hand in the central part of the gesture space
⇐	gesture in the left part of the gesture space
⇑	gesture in the frontal part of the gesture space
⇒	gesture in the right part of the gesture space
↗	gesture in the right-upper part of the gesture space
(r	right hand beating down with open palm
⋃	both hands beating down with open palms

While speaking, Simona continuously accompanies her words by moving her hands and arms. Within this sort of "gestural continuum," two striking features are seen: a peculiar use of pointings in her gesture space, and the recurrence of many beats, which accompany the development of the discourse.

Let us first consider the gesture space. The gesture space (McNeill, 1992) is the area in front of the speaker's body, in which he or she performs the majority of his or her gestures. For European languages, it roughly goes from the waist to the eyes and includes the space between the shoulders.[6] From the video analysis, but also from the pictures shown above, we can see that Simona organizes her gestures around four specific areas located in different regions with respect to her body: to the right, to the left, in the center, and related to the written sheet. Figure 4.14 shows the four areas.

Taking into account the co-timed words, we can correlate each area with a specific meaning and function. The areas around her body (shown as circles in Figure 4.14) have close relationships with the logical dimension of the argument. Specifically, we find three pointing gestures:

Figure 4.14 Four areas in Simona's gesture space: to the right, to the left, in the center, and related to the written sheet.

- to the *left* (coded ⇐), corresponding to the case in which Ulysses arrives after the cloth is finished (13d, Figure 4.7);
- to the *right* (coded ⇒), corresponding to the case in which Ulysses arrives before the cloth is finished (13f, Figure 4.9);
- in the *center* (coded ⇑), corresponding to logical connectives (13e [53"], Figure 4.8).

These gestures show pointing or deictic features. However, they do not point to anything present in the scene; they point to a location in space, an abstract space, and correspond to what McNeill (1992) calls *abstract pointing* (i.e., when there is no actual physical object that the speaker intends to reference by pointing). McNeill (1992) recognizes a certain meaning to such abstract pointing:

> the speaker appears to be pointing at empty space, but in fact the space is not empty; it is full of conceptual significance. Such abstract deixis implies a metaphoric use of space in which concepts are given spatial forms. (p. 173)

In our case, the two opposite (i.e., right and left) locations substantiate the two opposite aspects of the "conceptual significance" of Ulysses arriving either after or before Penelope completes the cloth. That is, they are related to the semantic content of the two cases of the argument, and their opposition in space expresses the logical opposition. The two opposed fields are linked by the gestures in the central space, which accompany and give emphasis to the logical connectives: a conjunction in line 13c ("and she," accompanied by a suspended left hand, coded —*l*) and a hypothetic-adversative preposition in line 13e ("and . . . if," with the abstract pointing ⇑).

Overall, the three ⇐, ⇒, ⇑ pointings together constitute a catchment whose function is to contribute to the structure of the argument from a logical point of view. A different function is accomplished by the several gestures pointing at the written sheet containing the text of the story, materially touched with the pen or hand. This type of pointing (coded ↓ in Table 4.1) is often co-timed with the protagonists of the story, either "Ulysses" (13a, b, i) or the "suitors" (13i). According to our interpretation, such gestures have the function of relating the argumentation to the narrative background provided by the story.

The semantic content of the logical argument is also accompanied by several beating gestures, which are performed with both hands (coded ⋃) or with the right hand (coded ⌊ *r*). In these cases, the palms are partially open upward and beat in time with the rhythm of her speech. Sometimes beats are performed through pointing gestures, as in the sequence accompanying line 13g (in Table 4.1, only one gesture is shown in Figure 4.10, but it actually repeats four times). In this case, the left hand also repeats an oscillating beat in the air, co-timed with the pointing made by the right hand. It is worthwhile to remark that the recurring features (catchments) of the shown gestures provide a structural cohesion to the whole complex sentence.

The last gesture, performed in the right-upper area of the gesture space and coded with ↗, has a hybrid function. It has both an analytic-structural function, because it marks the final consequence in the second part of the argument, and a semantic-narrative function, because the hand movement clearly indicates that the suitors must clear out. The semantic-narrative function is accentuated by the extended uttering of the final vowel "i" in the Italian word "pretendenti" (translated as "suitorssss"), with which the gesture is co-timed.

Integrating Gesture and Prosodic Analysis

In order to substantiate our claim about the analytic-structural function of gestures in the argumentation process, we integrate the gesture analysis developed above with the prosodic features of co-timed utterance. We report in Table 4.2:

- the original Italian and the English translation (preserving Italian word order);
- its prosodic intonation, indicated with <-------< for a rising intonation, and >-------> for a falling intonation;
- the codes for the co-timed gestures. The codes previously introduced (see Table 4.1) are used to match the stroke phases of gestures with the corresponding part of the utterance.

From the table, it becomes apparent that Simona is rhythmically accompanying her speech with her gestures. The constant presence of beating gestures underlines the effort that she is making in fully articulating the complex argument.

We have used the codes for gestures introduced in Table 4.1 and matched them with the words co-timed with the stroke parts of the gestures.[7] The utterance is written with accents indicating the stress of the Italian word: for example, in sentence b, "se venìva Ulìsse èntro cìnquànta giòrni." We can observe that there is a strong correlation between the beats ↓, (r, ⋃ and the stress of words, which is quite expectable.

Most interesting for our discussion is the correlation between the logical structure of the argument, the ⇒, ⇐, ⇑ catchments, and the intonation of the phrases. In the previous paragraph, we analysed the relationships between gestures and words, showing how the left and right areas of the gesture space acquire the "conceptual significance" (McNeill, 1992) of the

TABLE 4.2 Simona's Argument From Line 13 in Italian, with Timeline (in seconds)

↓ ↓ \r \r \r \r —l

Se veniva Ulìsse prìma… se veniva Ulìsse èntro cinquànta giòrni…

43 45 48

>--------------------------> >-->

If came Ulysses before… if came Ulysses before fifty days…

—l \r \r ⇐ ⇐

e lèi avèva già finìto di fàre la copèrta lèi sposàva un àltro.

49 51 53

<--< >---------------------->

and she had already completed the cloth she married another one.

⇑ V V ⇒ ⇒ ⇒ ⇒

E invèce se… lèi…lèi nòn finìva…lèi nòn finìva di fàre la copèrta èntro cinquànta giòrni

53 55 57 58 60

<------------< >-->

And on the contrary if… she…she did not finish…she did not finish to make the cloth within fifty days

↓ ↓ ↓ V V

lèi si prendèva di nuòvo Ulìsse… che èra già arrivàto

60 62 63 64

<--------------------------------------< >----------------------->

she took back Ulysses… who was already arrived…

↓ ↗

quìndi i pretendèntiiii… nòn potevàno più sposàrla

64 65 66 67 69

<-----------< >------------> <-------------------------------<

hence the suitorssss… could not marry her anymore.

two opposite cases and the central one a linking function between them. Now we can add that the two opposite cases are expressed through a raising intonation (>---->: 51"–53" and 55"–60") reaching a pitch (< >: 53"), whereas the logical connective is expressed in a falling intonation (<----<: 53"–55"). By coding rising intonation with <------< and falling intonation with >------>, it easy to locate the dipping or peaking points and observe how they mark the structural aspects of the speech. If we take into account these prosodic features and integrate them with the analysis of gestures, we obtain a new tree structure (see Figure 4.15).

First, we remark that the gesture-prosodic tree is compatible with the logical-linguistic one (shown in Figure 4.2). For instance, consider the stroke of the frontal gesture (Table 4.1, coded as ⇑). The stroke (54) marks the bifurcation of the main components of the tree, namely (43–53) versus

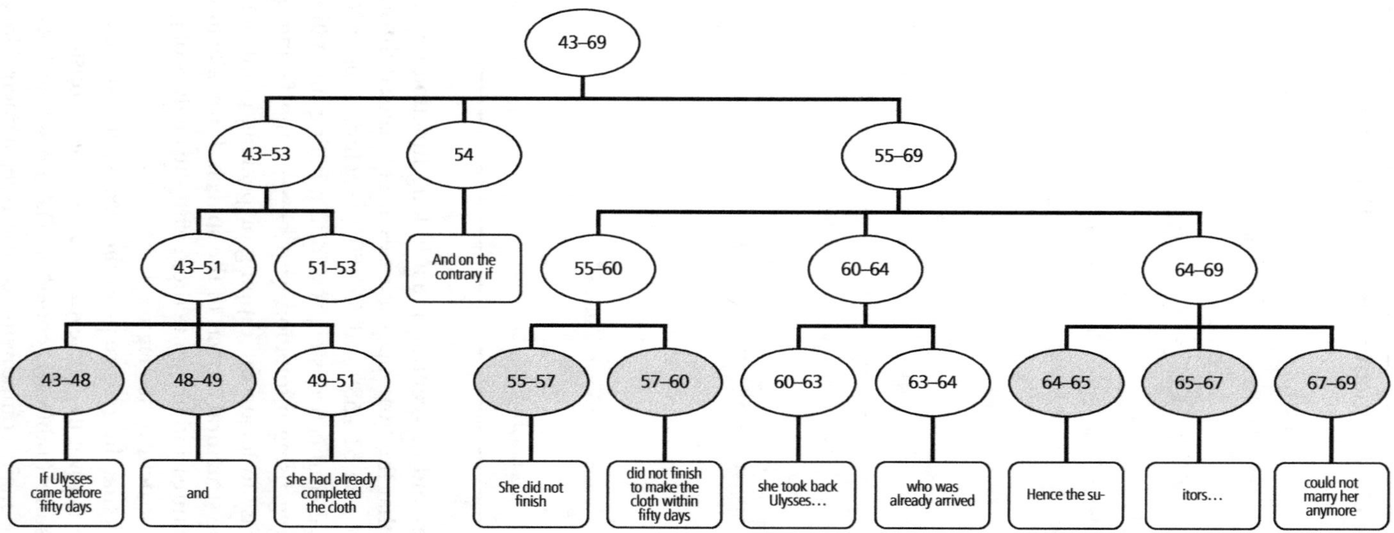

Figure 4.15 The tree structure of sentences 13 obtained with the integration of gesture and prosodic analysis (numbers refer to time in seconds). The shaded boxes indicate further specifications provided with the prosodic analysis.

(55–69), which are the two parts of the argument, namely, "Ulysses arrived after the cloth is completed, so..." versus "Ulysses arrives before the cloth is completed, so...." In the previous tree (Figure 4.2), obtained with the logic-linguistic analysis, the same function is absolved by the adversative sentence "And on the contrary," which is actually co-timed with the ⇑ gesture. However, we can say that the gesture-prosodic tree is more than compatible with the logical-linguistic tree; it is indeed a refinement of it. Comparing the two trees, we can see how the gesture-prosodic analysis results in sub-trees[8] of the previous one (in Figure 4.8, they are shaded). An example is given by the branching below the knot (64–69) into the three sub-knots (64–65), (65–67), and (67–69). In fact, looking at the utterances between 64 and 69, there is a peak in intonation at 65 ("hence the suitorssss") and a dipping at 67 ("could not marry her anymore"; see Table 4.2). These mark a change in the stream of Simona's semiotic productions. For this reason, we split them into three parts. This result further corroborates our hypotheses that the gestures in line 13 have analytic-structural functions and that they reflect the logical structure of Simona's argument.

As a final observation, it is important to point out that the analysis of gestures, integrated with the prosodic analysis, allows us to focus on details that are not grasped through a purely logical-linguistic analysis. For example, the sentence in 64–69 ("hence the suitorssss...could not marry her anymore") from a logical point of view is a consequence of the second case of Simona's argument. Using gestures it is possible to point out other aspects (e.g., the upsetting dispelling of the suitors), which are not explicitly stated in utterances. These gestures make visible with their imagistic function some aspects of the deep structure of the argument, which are missing if we consider only speech. In the final discussion, we will attempt to sketch how gestures are capable of expressing structural features of mathematical arguments.

DISCUSSION

In most literature on gesture, which is generally based on the analysis of everyday conversation, the global-holistic aspect of gestures is emphasized in contrast with the analytic character of language (McNeill, 1992; Sandler, 2009). Starting from these premises, in our research in the mathematics education domain, we have observed that gestures may also play specific roles in providing a logical structure to argumentation processes.

In this chapter, we have provided some evidence for this claim by analyzing in detail an episode from a primary school classroom. Specifically, considering an argument by cases that concludes the solution of a problem-solving activity, we have closely scrutinized the gesture-speech unity through which the argument is expressed. In our analysis, we have highlighted the following:

- The striking synergy between specific linguistic functions and the features of co-timed gestures. For example, the semantic content co-timed with pointings to the sheet containing the text of the story (↓) and logic connectives co-timed with abstract pointings (⇐, ⇒, ⇑).
- The common structure of the logical deep structure of an argument and that given by the different (falling vs. rising) intonations of utterances, which express it verbally.
- The role of catchments in providing the necessary cohesion for a complex argument. See the constant beats or the repeated features of the three abstract pointings (⇐, ⇒, ⇑).
- A specific use of the gesture space, which is represented by the striking example of the two right and left spaces, with their opposition in space expressing the logical opposition of the two cases.

Overall, we have documented a structural gesture-speech unity in mathematical argumentation (i.e., a gesture-speech unity that goes beyond the semantic content and involves deeply the structural aspects of argumentation). Our conviction is that such structural unity supports in relevant ways the coordination of semantic and logical aspects of argumentation.

This result may have interesting implications in the perspective of teaching and learning mathematical argumentation competences. It is well known that students of all ages meet great difficulties in facing argumentation and proving processes, where "structural" or "logical" understanding is needed. With these terms, we refer to the competences related to expressing and understanding ideas in those modalities that are established in our mathematical culture: from the simple use of implication ("if... then," "so," "hence") to proof by contradiction, by induction, and so on. In his studies in psychology, Bruner (1986) identifies this type of thinking based on cause-effect reasoning and producing logical arguments as "paradigmatic" or "logical-scientific." It is seen in opposition to "narrative thinking," which focuses on intentions and actions, and which is strongly anchored to experience in context. Often students' difficulties in mathematical argumentation can be related to the use of narrative thinking modalities and goals instead of logical-scientific ones. In fact, these two kinds of thinking (paradigmatic and narrative) represent the world in different ways and obey different laws, being seemingly irreducible to each other. One produces logical arguments whose goal is truth, whereas the other one produces stories whose goal is plausibility. So far, these difficulties have been studied mainly focusing on linguistic or symbolic productions. Shedding light on the role that gestures and nonsymbolic semiotic resources may play in structuring mathematical arguments (i.e., in the logical-scientific thinking), our research opens new directions for further study in a context—mathematical argumentation—where the supremacy of the discursive plane (at least in Western cultures) is still unquestioned.

Recent research has pursued these notions, stressing the core role that perception and action have in mathematical conceptualization (Cook & Goldin-Meadow, 2005; Nemirovsky, Rasmussen, Sweeney, & Wawro, 2011; Pozzer-Ardenghi & Roth, 2010). For example, Nemirovsky et al. (2011) write:

> Perceptuo-motor activities *incorporate* mathematics, making unnecessary the postulation of generic cognitive structures regulating perception and action from above, as it were. Mathematics is, accordingly, a way of perceiving and acting with respect to various symbols and notations resulting from complex historical traditions. These perceptuo-motor ways are expressed in and constituted by multiple bodily activities, such as talk, gesture, gaze, posture, writing, and so forth. (p. 2; italics original)

Our analytic-structural enquiry shows how such multiple body activities such as speech, gesture, and intonation constitute a compact core, which can express directly the so-called structural aspects of a mathematical argument. In a sense, we have reversed the usual reductionist step: We have not considered perception and action as a first step toward abstract structures in a sort of an input-output mechanism within a general information-processing system. On the contrary, we have shown how the embodied components, represented by the unit "speech-gesture," can incorporate and share the "structural" features of a mathematical discourse, encompassing all that we call abstract thinking.

ACKNOWLEDGMENTS

We wish to thank all the teachers working in our group of research without whose work and contribution this chapter could not exist, particularly Bruna Villa, in whose classroom the teaching experiment discussed in this chapter occurred. Warm thanks also to the students of her classroom, to their families who consented to their children being videotaped, and particularly to Simona, whose gestures brought us much to think about regarding gesticulation.

NOTES

1. The idiosyncratic character of gesture seems to have been overestimated: "It is often said that gesticulation is idiosyncratic, each speaker improvising his

own forms. So far as I know, no one has ever really tested this claim. My own experience in gesture-watching suggests to me that people are far more consistent in what they do gesturally than this 'idiosyncrasy' claim would lead one to imagine" (Kendon, 1996, p. 10).

2. The word comes from logic; in fact, the recursion notion pointed out in linguistics can be interpreted as a case of mathematical recursion, namely, as a function that applies to itself (see Chomsky, 1965).
3. For example, a contribution on the role of rhythm in algebraic generalization is found in Radford, Bardini, and Sabena (2006).
4. Limitations of space do not allow us to develop the role of the narrative plane in problem solving. The reader may find a brief account in Bazzini, Saben, and Villa (2009).
5. According to the text of the story (which is always kept in the shared space in front of the children), Ulysses was arriving at Ithaca in exactly 50 days. The children probably misread this temporal specification and transformed it into "before 50 days." The original problem "Does Penelope finish the cloth before 50 days?" is easily transformed into "Does Penelope finish the cloth before Ulysses arrives?" This kind of phenomenon has been described in problem-solving literature by R. Zan (2011), who, quoting U. Eco, speaks of children as losing themselves in the narrative wood. However, in our case, this added variable does not prevent the students from solving the mathematical problem.
6. However, differences have been detected according to cultural setting and age (with children the space is larger).
7. For this reason, we did not underline the text in this table.
8. For the notions of tree and sub-tree, see Diestel (2005).

REFERENCES

Arzarello, F., Bazzini, L., Ferrara, F., Robutti, O., Sabena, C., & Villa, B. (2006). Will Penelope choose another bridegroom? Looking for an answer through signs. In J. Novotná, H. Moraová, M. Krátká, & N. Stehlíková (Eds.), *Proceedings of the 30th Conference of the International Group for the Psychology of Mathematics Education* (Vol. 2, pp. 73–80). Prague, Czech Republic: Charles University Press.

Arzarello, F., Paola, D., Robutti, O., & Sabena, C. (2009). Gestures as semiotic resources in the mathematics classroom. *Educational Studies in Mathematics, 70*(2), 97–109.

Bara, B. G., & Tirassa, M. (1999). A mentalist framework for linguistic and extra-linguistic communication. In S. Bagnara (Ed.), *Proceedings of the 3rd European Conference on Cognitive Science* (pp. 285–290). Rome, Italy: Instituto di Psicologia, CNR.

Bazzini, L., Sabena, C., & Villa, B. (2009). Meaningful context in mathematical problem solving: A case study. *Proceedings of the 3rd International Conference on Science and Mathematics Education* (pp. 343–351). Penang, Malaysia: CoSMEd.

Bruner, J. (1986). *Actual minds, possible worlds.* Cambridge, MA: Harvard University Press.

Chomsky, N. (1965). *Aspects of the theory of syntax.* Cambridge, MA: MIT Press.

Cook, S. W., & Goldin-Meadow, S. (2005). The role of gesture in learning: Do children use their hands to change their minds? *Journal of Cognition and Development, 7*(2), 211–232.

Diestel, R. (2005). *Graph theory* (3rd ed.). Berlin, Germany: Springer-Verlag.

Donaldson, M. (1978/2006). *Children's minds.* New York, NY: HarperCollins.

Edwards, L. D. (2009). Gestures and conceptual integration in mathematical talk. *Educational Studies in Mathematics, 70*(2), 127–141.

Goldin-Meadow, S. (2003). *Hearing gestures: How our hands help us think.* Cambridge, MA: Harvard University Press.

Hauser, M. D., Chomsky, N., & Fitch, W. T. (2002). The faculty of language: What is it, who has it, and how did it evolve? *Science, 298*(5598), 1569–1579.

Hirst, D., & Di Cristo, A. (Eds.). (1998). *Intonational systems.* Cambridge, UK: Cambridge University Press.

Kendon, A. (1996). An agenda for gesture studies. *Semiotic Review of Books, 7*(3), 8–22.

McNeill, D. (1992). *Hand and mind: What gestures reveal about thought.* Chicago, IL: University of Chicago Press.

McNeill, D. (2005). *Gesture and thought.* Chicago, IL: University of Chicago Press.

McNeill, D., Quek, F., McCullough, K.-E., Duncan, S., Furuyama, N., Bryll, R., & Ansari, R. (2001). Catchments, prosody and discourse. *Gesture, 1*(1), 9–33.

Nemirovsky, R., Rasmussen, C., Sweeney, G., & Wawro, M. (2011). When the classroom floor becomes the complex plane: Addition and multiplication as ways of bodily navigation. *Journal of the Learning Sciences, 21*(2), 287–323.

Pozzer-Ardenghi, L., & Roth, W.-M. (2008). Catchments, growth points and the iterability of signs in classroom communication. *Semiotica, 172*(1/4), 389–409.

Pozzer-Ardenghi, L., & Roth, W.-M. (2010). *Staging & performing scientific concepts: Lecturing is thinking with hands, eyes, body, & signs.* Rotterdam, the Netherlands: Sense Publishers.

Radford, L. (2010). The eye as a theoretician: Seeing structures in generalizing activities. *For the Learning of Mathematics, 30*(2), 2–7.

Radford, L., Bardini, C., & Sabena, C. (2006). Rhythm and the grasping of the general. In J. Novotná, H. Moraová, M. Krátká, & N. Stehlíková (Eds.), *Proceedings of the 30th Conference of the International Group for the Psychology of Mathematics Education* (Vol. 2, pp. 393–400). Prague, Czech Republic: Charles University.

Sandler, W. (2009). Symbiotic symbolization by hand and mouth in sign language. *Semiotica, 174*(1/4), 241–275.

Saussure, F. (1916). *Cours de linguistique générale.* Paris, France: Payot.

Zan, R. (2011). The crucial role of narrative thought in understanding story problems. In K. Kislenko (Ed.), *Current State of Research on Mathematical Beliefs XVI, Proceedings of the MAVI-16 Conference* (pp. 287–305). Tallinn, Estonia: Institute of Mathematics and Natural Sciences, Tallinn University.

CHAPTER 5

AN EXPLORATORY STUDY OF MULTIMODALITIES IN THE MATHEMATICS CLASSROOM

Enrica's Explanation

Francesca Ferrara
Ornella Robutti
Università di Torino, Italy

Laurie D. Edwards
Saint Mary's College of California

In this chapter, we focus on the multiple modalities in play in the classroom when students work with mathematical meanings. In the mathematics classroom, it is typical to see students and teachers utilizing oral speech, written text and symbols, and graphical inscriptions, all in the service of constructing and communicating about mathematical ideas. In the episode analyzed here, a student utilized two primary modalities—speech and gesture—to propose and explain her ideas about solving a mathematical problem posed to the class as a whole. The analysis of this student's multimodal communication

Emerging Perspectives on Gesture and Embodiment in Mathematics, pages 105–124

can help shed light on the complementarity and richness of different modalities in the construction and expression of mathematical thinking.

In the episode being analyzed, a class of 10-year-old students explores the process of measuring distance through the use of a technological artifact, the pedometer, whose use had been demonstrated to the class by the researcher. The interaction with the tool requires students to introduce an input—specifically, the length of the average step. During a class discussion, it becomes clear that there needs to be an agreement on how to find this length. This moment is crucial to the understanding of the experience by the entire class, and one student, Enrica, volunteers a solution as well as a detailed explanation of this solution. We focus here on Enrica's explanation because of its richness in terms of the multiple modalities in play. By multiple modalities, we refer to the various embodied means that are used by the teacher and the students in mathematics teaching and learning, including gesture, speech, bodily actions, writing, facial expression, and tone of voice.

To frame our analysis, we begin by presenting research on multimodality and gesture from the fields of psychology and mathematics education. Then we outline a view of multimodality with respect to the learning of mathematics, and finally we present, analyze, and discuss the classroom episode in detail.

MULTIMODALITY

Over the last decade, the notion of multimodality has been considered from different perspectives in different scholarly fields. In the field of communication and semiotics, multimodality deals with all the means we have for making meanings (i.e., the various modes of representation considered in terms of their specific ways of configuring the world; Kress, 2005; Kress, Charalampos, Jewitt, & Ogborn, 2006). In a departure from earlier studies, which generally discussed only the modality of language, Kress and his colleagues broaden the traditionally language-focused view of knowledge in order to examine the interface between language and other communicative resources. For example, in a study of science teaching, they attend also to "extralinguistic stuff," noting that:

> Language is not absent from our discussion, but nor is it central. It is, as we have said, "there," but "there" among all other modes: foregrounded when it is, in the representational ensemble, and not foregrounded in our descriptions when it is not so in the data. The so-called extralinguistic stuff is no longer a residual category, peripheral to the analytic act of describing the phenomenon under study. It reminds us that... classrooms should be seen as a cultural setting with their own semiotic systems where certain types of

> knowledge are produced, distributed and reinforced. (Kress, Charalampos, Jewitt, & Ogborn, 2006, p. xii)

Working within this perspective, Kress (2000) develops the idea that different modalities offer different capabilities or affordances for communication. Indeed, Kress (2000) claims that:

> The semiotic modes of writing and of image are distinct in what they permit, that is, in their *affordances*. Image is founded on *the logic of display in space*; writing (and speech even more so) is founded on *the logic of succession in time*. Image is spatial and nonsequential; writing and speech are temporal and sequential. That is a profound difference, and its consequences for representation and communication are now beginning to emerge in this semiotic revolution. (p. 339)

Kress (2001) makes a distinction between modes (ways of communicating meanings, such as language, images, and sound) and media (the material means for expressing these modes). Among the latter, he would include pen and paper, stone and chisel, and the human vocal apparatus (Kress, 2001). He does not explicitly include the human voice or other bodily resources, such as gesture, as modes; rather, they are considered only as possible media for communication. Instead, we treat multimodality in a broader sense that would include gesture and other bodily movements and states as modes with their own affordances and constraints. From this perspective, multimodality encompasses all the cultural, social, and bodily resources available not only for communicating but also for reasoning. In so doing, we draw on recent work in neuroscience and linguistics on multimodality (Gallese & Lakoff, 2005; see Chapter 1, this volume). Among the multiple modalities used by humans, one of the most studied, particularly in the fields of psychology and linguistics, is that of gesture.

GESTURE

Gesture has been extensively studied in the last 20 years, especially in terms of its relationship with speech and thought (Alibali, Kita, & Young, 2000; Bavelas, Chovil, Coates, & Roe, 1995; Goldin-Meadow, 2003; Kendon, 2004; McNeill, 1992, 2005; Radford, 2009). In his book *Hand and Mind*, McNeill (1992) defined gestures as "movements of the arms and hands" that "are closely synchronized with the flow of speech" (p. 11). McNeill (1992) emphasizes the link between gestures and internal imagery, stating that gestures, "display inner thoughts and ways of understanding events of the world... they belong, not to the outside world, but to the inside one of memory, thought, and mental images" (p. 12).

In his investigation of the interplay between gestures and thought, McNeill (1992) distinguished different categories of gestures according to their specific function with respect to speech. McNeill labeled these function-based categories iconic, metaphoric, deictic, and beats. The iconic and metaphoric categories are similar in the sense that each involves gestures with imagistic or pictorial content. However, iconic gestures "bear a close formal relationship to the semantic content of speech" (McNeill, 1992, p. 12), whereas in a metaphorical gesture, the pictorial content provides a source domain for a conceptual mapping to more abstract, nonimagistic content. Deictic gestures are familiar pointing gestures (often accompanying indexical words such as "there" or "this"), which indicate objects or events with or without physical existence in the world. Beats use repeated "one-two" movements, follow the rhythm of speech, and index an accompanying word or phrase "as being significant, not for its own semantic content, but for its discourse-pragmatic content" (p. 15). In more recent work, McNeill (2005) stresses the necessity of speaking in terms of dimensions instead of categories because a single gesture may have loadings across different dimensions rather than belonging strictly within one category. In this chapter, we embrace this dimensional perspective to study the continuum of thinking processes, examining not only the type of gesture but also its semiotic power with respect to the construction of knowledge (Arzarello, Paola, Robutti, & Sabena, 2009; Ferrara, 2006). In other words, our interest is in the complex interplay between different modalities, including gestures and speech, in constructing and communicating mathematical ideas.

Researchers in psychology and linguistics, such as Alibali, Kita, and Young (2000) and Goldin-Meadow (2003), have highlighted relations between gesture and the spoken word and, in particular, have discovered specific affordances of gesture in learning situations. Both have found that if the information content of a gesture differs from that of the speech, this "mismatch" (Goldin-Meadow, 2003) or "non-redundancy" (Alibali, Kita, & Young, 2000) can indicate that the learner is on the verge of reaching a new understanding. In this situation, the gesture appears to afford the learner the opportunity to express new knowledge in advance of his or her ability to express it in words.

Thus, gesture is not simply (or not always) a form of illustration accompanying speech but can play an important role in building new knowledge. Several studies in mathematics education research have highlighted this role with respect to the construction of mathematical meanings (e.g., Arzarello & Robutti, 2008; Edwards, 2008; Ferrara, 2006; Nemirovsky, 2003; Radford, 2009; Robutti, 2006).

MULTIMODALITY IN LEARNING MATHEMATICS

In 2009, a special issue of *Educational Studies in Mathematics* was devoted to the study of gestures in mathematics learning. This special issue used the perspective of multimodality to examine the construction of mathematical meanings, with specific attention to the use of gesture by teacher and students (e.g., Arzarello et al., 2009; Nemirovsky & Ferrara, 2009; Radford, 2009). From this perspective, multimodality in mathematics was conceived as being "the range of cognitive, physical, and perceptual resources that people utilize when working with mathematical ideas. These resources or modalities include both oral and written symbolic communication as well as drawing, gesture, the manipulation of physical and electronic artifacts, and various kinds of bodily motion" (Radford, Edwards, & Arzarello, 2009, pp. 91–92). A central thesis was that cognition has a "multimodal nature," meaning that, "in our acts of knowing, different sensorial modalities—tactile, perceptual, kinesthetic, etc.—become integral parts of our cognitive processes" (Radford et al., 2009, p. 92).

The thesis that mathematical cognition is multimodal arises not only from the results of psychological studies on gesture but also from research in mathematics education that draws attention to embodiment and perceptuo-motor activity in mathematics learning. For example, Nemirovsky (2003) states that:

> [T]he understanding of a mathematical concept rather than having a definitional essence, spans diverse perceptuo-motor activities which become more or less significant depending on the circumstances. For instance, seeing a trigonometrical function as a component of circular motion or as an infinite sum of powers may entail distinct and separate perceptuo-motor activities. Learning a different approach for what appears to be the "same" idea, far from being redundant, often calls for recruiting entirely different perceptuo-motor resources. (p. 108)

Speech, gesture, and other bodily movements are among the various modalities utilized in the perceptuo-motor activities involved in learning mathematics, whether a student is "simply" writing down an equation or interacting with a device that tracks and records her distance from a sensor. Learners use these modalities in an integrated way, not one at a time. These modalities are particularly important for the analysis of students' cognitive processes because they provide a broader source of information than do speech or written inscriptions alone. For example, if a student makes an iconic gesture of a straight line in the air while discussing the solution of a problem, this may indicate that she has the intuition that the solution is a straight line, and the straight line has a specific representation in a Cartesian plane. This understanding may or may not yet be expressed through the symbolic inscription of an equation or a numeric representation in a table of two variables.

Of course, all symbolic representations within mathematics are the product of a long-term process of cultural evolution and are shared among the community of mathematicians. However, the understanding and representations given by a student could differ from the "institutional" one. Aligning the individual meanings with the cultural meaning of the community is the aim of the teaching/learning process, and the role that modalities play in these processes is fundamental because they act as a "bridge" between the personal, individual meanings, and the cultural, historically framed meanings of the community of mathematicians. Theoretical perspectives based on semiotics have investigated the bridging of the gap between individual and cultural meanings in mathematics (Arzarello & Robutti, 2008; Radford, 2008; Robutti, 2006). Bringing in the analysis of the modalities used by students and teachers in the classroom provides researchers more access to the processes involved in connecting individual to cultural meanings.

For example, a teacher might present a gesture with two upraised palms held at the same height while discussing "balancing" equations, reinforcing this metaphor visually. A student might introduce a curve in the air using a hand to represent an accelerated motion or span a set of symbols on paper with his hand to indicate a portion of an equation. These modalities are not culturally accepted representations of mathematical objects but help students and teachers in constructing meanings for them. They can act as first steps in helping students to generalize, specialize, and understand more abstract definitions, concepts, and theorems during the process of constructing meanings.

In addition, each modality has particular affordances and characteristics that become more or less apparent depending on the circumstances and their use in context. For example, gestures or other bodily actions can isolate one portion of an inscription, particularly when using deictic gestures. Prosody in oral speech can give a rhythm to a sequence of words (e.g., a choral practice of multiplication facts), assisting in their later retrieval. A change of volume or tone of voice can be used to highlight a specific word or algebraic expression. Pauses in speech can allow students time to absorb new information. A gesture made with a hand or an artifact (pen, pencil) can reveal how a learner conceptualizes the "shape" of a mathematical concept. The affordances of the different modalities are what allow students to bridge the gap, in small steps, as they begin to associate their individual meanings to the meanings of the mathematical community.

Illustration: Enrica's Explanation

As an illustration of the framework of multimodality and affordances, we present an episode from a study of the cognitive strategies of Grade 5 primary school students (ages 10–11), who were asked to describe and interpret graphs and tables related to uniform motions. In the learning activity, they performed the uniform motion by walking straight paths between different locations. This activity was part of a teaching experiment about direct proportionality between quantities, in this case, between space and time. This concept was explored and modelled in various environments: numerical, graphical, and symbolic, and with both paper and pencil as well as technological devices (a pedometer and a chronometer to collect data in the experiment, the blackboard to draw the representation of motion, and a graphing calculator to construct a model of the motion).

In the following excerpts, the students, who are at the beginning of the teaching experiment and have seen the researcher using the pedometer, are engaged in a collective discussion coordinated by the researcher about how the pedometer works. In fact, the pedometer works by taking as input the step length of the walking subject and giving as output the total distance covered—having counted the steps made by the walking subject. After the researcher has explained this, the students have to choose which data are to be inserted as input into the pedometer in order to use it for measuring a distance of about two kilometers from school to the library. The correct answer would be to use the average step of the subject in order to obtain the most accurate distance as output.

The students first discuss how they can measure the size of a step, and they decide to measure the distance from the tip of one shoe to the tip of the following shoe for the distance "one step" (calling it the tip-to-tip technique). Then measuring the distance from the first tip to the last tip of a sequence of steps would give them the total distance covered. Dividing that distance by the number of steps in the sequence, they obtain the average length of one step of the moving subject.

The discussion presented in the example below takes place in the class, as the students are deciding which data to insert in the pedometer. First, the researcher takes three steps, each intentionally different in length from the others, to make the students aware of the fact that they should consider an average value rather than simply choosing an arbitrary step to measure. Then the students discuss how to calculate this input value. One student in particular, whom we have called "Enrica," has clear ideas on how to do it. In the excerpt below, she explains her ideas to her classmates, the teacher, and the researcher:

Enrica: You have made three steps, so one of 54, one of 57, one of 75 centimeters, you made them different, then we sum them up and we do the average value and we obtain, as average, the length of the steps.

Researcher: That's a good idea.

Antonella: You take a step and then you multiply by three.

Researcher: I take a step and then I multiply it by three, yes, this is another idea. And there is one method more... say, can we try to make all the steps equal?

Someone: No!

Enrica: We can try.

Researcher: [...] so I suggest you to try to make some measurements in every group, of your steps when walking and then we can use Enrica's idea of... what did she tell us, Sara?

Sara: To do the average value.

In Table 5.1, we present Enrica's verbal explanation again, showing its interplay with the modalities of gesture and head movement. The words spoken during the stroke (or most emphatic) phase of each gesture are underlined.

TABLE 5.1 Enrica's Explanation

Speech	Gestures	Head Motion
Enrica (on the right): *You have made <u>three steps</u>* (Figure 5.1)	 Figure 5.1: *"three steps"* (left hand perpendicular to the desk, fingers slightly spread)	Head moves forward (chin closer to chest)
Additional gestures: Turning the palm of the left hand up and starting to move it up		

(continued)

TABLE 5.1 Enrica's Explanation (continued)

Speech	Gestures	Head Motion
so one (Figure 5.2a)	 Figure 5.2a: "*one*" (moving her open hand up)	Head lifted up
of (Figure 5.2b)	 Figure 5.2b: "*of*" (moving her open hand down)	Head up, starting to move forward (chin farther from chest)
54 (Figure 5.2c)	 Figure 5.2c: "*54*" (moving her open hand to the notebook)	Head moves forward (chin closer to chest)

Additional gestures: For each number, she repeats the same movements with the hand and the head shown in Figure 5.2a, b, and c (lifting the left hand with palm facing up, lowering the hand, and then touching the notebook with the back of the hand)

(continued)

TABLE 5.1 Enrica's Explanation (continued)

Speech	Gestures	Head Motion
one of *<u>57</u>* (Figure 5.3)	 Figure 5.3: "*57*" (again, putting her open hand on the notebook)	Head moves forward (chin closer to chest)
Additional gestures: At the word "57," the left hand moves with the palm up, but suddenly the palm turns down		
one of 75 *<u>centimeters</u>* (Figure 5.4a)	 Figure 5.4a: "*centimeters*" (touching the notebook with her open hand)	Head stops in forward position (chin almost touching chest)
Additional gestures: At the word "centimeters," the head (moving down) stops, and the left palm turns down		

(continued)

TABLE 5.1 Enrica's Explanation (continued)

Speech	Gestures	Head Motion
you made them <u>different</u> (Figure 5.4b)	 Figure 5.4b: "*different*" (hand palm up, fingers slightly open toward outside)	Head stops in forward position (chin almost touching chest)
Additional gestures: At the words "you made," the left palm turns up		
Enrica begins to use two hands		
then <u>we sum them up</u> (Figure 5.5)	 Figure 5.5: "*we sum them up*" (brings fingertips of both hands together, palms facing chest, and hands perpendicular to the desk)	Head up, starting to move forward

(continued)

TABLE 5.1 Enrica's Explanation (continued)

Speech	Gestures	Head Motion
and we do the average (Figure 5.6)	 Figure 5.6: "*average*" (opens both hands with fingers arched back, moves hands and arms outward, still just above the desk)	Head stops in forward position
And we obtain, as (Figure 5.7)	 Figure 5.7: "*as*" (left palm horizontal, 60° to desk, right fingers cover left palm)	Head lifted up
average (Figure 5.8)	 Figure 5.8: "*average*" (repeats gesture in Figure 5.6)	Head moves forward, stops in forward position

(continued)

TABLE 5.1 Enrica's Explanation (continued)

Speech	Gestures	Head Motion
the length (Figure 5.9)	 Figure 5.9: "*length*" (brings finger tips of both hands together, palms still apart)	Head held up
of the steps (Figure 5.10)	 Figure 5.10: "*steps*" (opens her hands with palms perpendicular to desk and facing each other about shoulder width apart)	Head held up

Multimodal Analysis of Enrica's Explanation

In terms of the mathematics of the activity, Enrica immediately proposes a solution: to add the lengths of each of the three steps taken by the researcher to determine the average step size. This average is the data to be used as input to the pedometer; once this is set, the students can use it to measure distances covered by walking.

Enrica's explanation of her solution is rich with multimodality. The gestures that accompany Enrica's speech and her tone of voice are meaningful: At almost every word, there is a gesture with a particular shape, function, intentionality, and affordance. It seems that the student is shaping through gestures the mathematics useful for her activity. In addition to the gestures that accompany speech, Enrica also uses head movements and voice intonation as modalities in constructing and expressing her knowledge.

She begins by referring to the three steps taken by the researcher when she introduced the problem of finding an input for the pedometer. As Enrica states the numerical measurement of each step, she uses a beating gesture with the left hand moving down to the desk—the palm up—with a rhythm insisting on the measures of the three steps, with the hand going down at every stated measurement (Figures 5.2a, 5.2b, 5.2c, and 5.3). The measurements' unit, centimeters, is then underlined by a palm-down gesture that closes the previous series (Figure 5.4a).

Enrica then depicts the operation of the sum of these three measures by bringing both hands together, palms inward and tips of corresponding fingers touching (Figure 5.5). In expressing the idea of addition, this gesture has a metaphorical dimension, expressed through the iconicity of physically "bringing together" or collecting objects. In these terms, it recalls the "Arithmetic Is Object Collection" metaphor, one of the basic metaphors of arithmetic identified by Lakoff and Núñez (2000).

She next offers a gesture coordinated with her words about finding the average, opening her arms and hands, palms up (Figure 5.6). This gesture, rather than being an attempt to communicate the idea of mathematical average, is more likely an interactive gesture (Bavelas et al., 1995). It belongs in a family of gestures used to metaphorically "offer" an idea to interlocutors. The simplest of this type of gesture is putting the palm up and toward the interlocutor, as if offering an object. These gestures are based on the metaphor that communication is a transfer of a material object, a ubiquitous though culturally specific metaphor (Bavelas et al., 1995; McNeill, 1992). One interpretation is that at this point, Enrica is "offering" the idea of using the average to her audience; the two-handed, open gesture also carries the suggestion (again culturally specific) that this idea should be accepted as obvious. In this case, the gesture also conveys the meaning generated by the whole process of explaining how to arrive at the idea of average; it is a meta-level gesture indicating that the solution and explanation offered so far are being made available to all those listening. This is reinforced by the posture of the girl, projected toward the interlocutor (head forward and shoulders up a little) (Figure 5.6) and by the repetition of the same word ("average") with the same gesture (Figures 5.6 and 5.8). Finally, once the average value is introduced and shared within the class, Enrica proposes that the average should be used to find the length of the steps, with "length" indicated by bringing together the fingertips of the two hands (Figure 5.9), this length to be used as input for the pedometer. The segment ends with Enrica saying "of the steps" while holding her two hands apart and facing each other.

This final gesture is interesting because it is commonly used to indicate "length," but here it occurs in conjunction with the phrase "steps." However, the entire phrase that Enrica utters is "length of steps," and it is conceivable that the referent of the final gesture is the entire phrase. Alternatively, she may have added the final gesture to reinforce the notion of the length of the steps (rather than their quantity or some other characteristic).

Two Phases of Gestures

The expressivity of Enrica's gestures is high, and, as noted earlier, she uses gesturing continuously in conjunction with her speech. However, a careful analysis reveals that the gestures displayed during this segment have two different kinds of referents depending on the phase and nature of the discourse. The beating and iconic gestures at the beginning of the sequence characterise the process of movement (taking steps, measuring, and counting), and generally only one hand is used. In contrast, the metaphorical gestures that follow involve both hands and include a symmetric opening and closing of the two hands. These gestures either refer to mathematical objects (sum, average, length), rather than physical actions, or they are interactive gestures of communication.

During the first group of gestures, Enrica is recalling the action of the researcher as she took the three steps and measured them, a phenomenological experience. However, the second group of gestures are directed to the construction of meanings that are new at that moment for the students and whose nature is mathematical or abstract rather than phenomenologically concrete. With these latter gestures, Enrica moves toward abstraction and, using both speech and gesture, proposes the concept of the average step length to express the appropriate input for the pedometer.

Multimodal Expressivity

The expressivity shown by this student is not only based on gestures; along with gestures, she coordinates head movements with her speech, moving her head down when she wants to underline a word or concept, up when preparing that movement or emphasizing a point. Except for the initial "beating" gestures, Enrica uses her hands in a symmetric and synchronised way. Gestures, words, and head movements generally match each other in this brief excerpt; this redundancy across modalities suggests that Enrica fully understands the ideas that she wants to depict and explain to the teacher and her classmates.

We can call Enrica's use of different modalities simultaneously multimodal expressivity; as noted earlier, this kind of expressivity is typical in human communication. Each modality has different characteristics and affordances: the speech is linear and analytic; the gestures and the head movements are global and synthetic. Furthermore, their diverse characteristics let them overlap and integrate with each other to express the same meaning. For example, as Enrica orally states the lengths of the three steps made by the researcher (Figures 5.1, 5.2a, 5.2b, 5.2c, and 5.3), she also uses her left hand to beat each measurement, and she simultaneously moves her head down. These three modalities used together demonstrate Enrica's intentionality of emphasizing the steps and their measures to set up and capture the audience's attention for the next part of the discourse: describing how to calculate the average value of the steps. These three modalities overlap each other, but each one of them has its particular affordances. The words (speech) provide the exact measurements of the steps; the gestures give emphasis to the numerical values attached to each measurement, marking the discreteness of each step; while the head amplifies this emphasis.

Gesture not only gives emphasis to the words, but the particular gestures this student uses also mirror and highlight certain specific aspects of the physical steps that Enrica refers to in her speech: that the leg and the person stand perpendicular to the ground, that the steps follow each other roughly in a straight line, and that they occur rhythmically. The numbers stated in Enrica's speech highlight the fact that the length of each step is not the same, an important consideration in solving the problem at hand. In this situation, both modalities correspond to the general meanings of measuring physical footsteps, but they highlight different aspects: the numerical length, in the case of speech, and their physical nature as well as distinctiveness, in the case of gesture.

After stating the final measurement, 75, in the next instant, Enrica makes use of the word "centimeters" (Figure 5.4a), and she accompanies it with a new gesture, placing her left hand palm down on the table. This gesture, we propose, is different in form from the previous ones because it has a different status in the process Enrica is explaining. It is used to convey the unit of measure (centimeter), which is important because it needs to be used when inserting the value in the pedometer. Her head also moves down at this point, emphasizing the word "centimeters."

In the second group of gestures, when Enrica describes the process used to find the average value, her multimodal expressivity is shown by gestures with both an iconic and a metaphorical dimension. Iconicity can be seen in the shape of the gesture of bringing together the two hands (Figure 5.5), which accompanies the words, "we sum them up." This "gathering" gesture is iconic to the corresponding physical action of bringing objects together but also metaphorical to the mathematical operation of addition. The subsequent gesture in which Enrica opens her hands and arms to "offer" the solution to the audience (Figure 5.6) is also metaphorical; as discussed above, it is a good example of the metaphor of communication as the transfer of a material object or substance.

DISCUSSION

The analysis of Enrica's episode shows the complex dynamic interplay of multimodalities and how they intertwine to bridge the gap between the students' personal senses (stimulated by the use of the tools) and the mathematical knowledge related to the concept of average step length. Each of the modalities Enrica utilizes in expressing herself participates in the communication process in a complementary way while "conveying" the same general information. This multimodality is demonstrated by the simultaneous talking, gesturing, and motions of the head, as well as by posture and other bodily expressions not analyzed here. Each of these modalities has a different affordance and plays a different role in the communication process. At the beginning of the episode, Enrica's words have the specific function of recalling the walking experience; the numbers she states allow the specification of the three steps in terms of their lengths. The language used is linear and analytic and draws on semiotic systems well known to her interlocutors (spoken Italian and whole numbers). At this point, Enrica is expressing mostly the phenomenological side of the experience: the taking of steps and measuring them. The use of "you" as a subject in her speech is emblematic in this moment. In this phase, the accompanying gestures and head movements mainly have two different functions: On the one hand, they serve to give emphasis to specific words; on the other hand, the gestures provide a visual corollary for the concepts recalled by those words (e.g., the measures of the three steps). For example, Enrica's left hand is always on the notebook, turned face down, when she recalls the length of a step. In a similar vein, the head stops in a position close to the chest each time Enrica refers to these measurements. The repetition of the same gestures and the same head movements highlights the importance (and similar role) of these measurements and also gives rhythm to the speech that helps her classmates follow Enrica's reasoning.

In the second phase of the episode, again, the character of the spoken language is linear and analytic, while that of head movements and gestures is global and synthetic; each modality reinforces the meaning expressed in the other. For example, when discussing the sum of the measurements, Enrica brings her hands together while leaving an open space between her arms. This global-synthetic gesture provides an instantaneous image of an idea that is communicated linearly and one piece at a time via speech. During this phase, we can also note a change of perspective, in which Enrica moves from the phenomenological side to focus more on the mathematics. In addition to evoking the sum of the step lengths, she also proposes the use of the average as the input value for the pedometer. While the phenomenological side was linked to the use of "you," now the subject in the speech becomes "we"; that is, the community of students and teacher are put at the center of the thinking process. The change is also evident in the new gestures that enter the scene. Instead of beats and iconic gestures, we now see gestures with a metaphorical dimension (e.g., for the sum, the average value, etc.) and an abstract reference.

The complementarity of the modalities in play can be appreciated through a thought experiment. If, while watching the video of this brief class episode, we turned off the audio (speech), we would be able to identify the appearances of many gestures and head movements. We would be able to grasp that some gestures are repetitive and some head movements appeared to be used as a means to emphasize certain elements of the discourse. But we would not be able to understand the full meaning or depth and richness of Enrica's reasoning without access to the modality of speech. A similar argument may be used with respect to the affordances of the other modalities. For example, if we turned off the video and listened only to the audio, we would lose the visual markers of the intensity of the communication, as well as the imagery displayed in Enrica's different gestures. Thus, it is important to be aware that single modalities, analyzed in isolation, cannot provide a full picture of what is going on within the classroom. The goal, when engaging in an analysis of mathematical activity from the perspective of multimodality, is to build an understanding of learning, teaching, and thinking that both utilizes the various kinds of information communicated by all the modalities present and also takes account of their roles in the construction of the knowledge.

In this chapter, we have proposed that physical action and sensory capabilities form a foundation for constructing mathematics, which is embodied both directly and indirectly via complex cognitive activity that spans multiple modalities. We see mathematical practice, like other forms of intellectual work, as utilizing multiple modalities, from basic sensory channels and motor actions through more abstract semiotic modes such as language and imagery. The affordances of the modalities in play were considered in a brief episode from a classroom task involving both physical action and mathematical reflection. What is relevant in this perspective is the interplay among the different modalities because each modality—when employed—prompts meaning via its particular affordances and adds force and emphasis to the others. In our case, this happens thanks to the intertwinement of gestures, language, and head movements, which sustain the effectiveness of one another.

In analyzing multimodality, we want to go beyond distinguishing categories of gestures or identifying whether there is redundancy between modalities. Instead, we want to examine the range of modalities, with their dimensions and functions (and, so, their affordances) within the complex landscape of mathematical learning, where personal sensory and motor experience must be linked to the mathematics developed by the community to explain situations and solve problems.

REFERENCES

Alibali, M., Kita, S., & Young, A. (2000). Gesture and the process of speech production: We think, therefore we gesture. *Language and Cognitive Processes, 15*(6), 593–613.

Arzarello, F., Paola, D., Robutti, O., & Sabena, C. (2009). Gestures as semiotic resources in the mathematics classroom. *Educational Studies in Mathematics, 70,* 97–109.

Arzarello, F., & Robutti, O. (2008). Framing the embodied mind approach within a multimodal paradigm. In L. English (Ed.), *Handbook of international research in mathematics education* (pp. 720–749). New York, NY: Taylor & Francis.

Bavelas, J. B., Chovil, N., Coates, L., & Roe, L. (1995). Gestures specialized for dialogue. *Personality and Social Psychology Bulletin, 21,* 394–405.

Edwards, L. D. (2008). Conceptual integration, gesture, and mathematics. In O. Figueras & A. Sepúlveda (Eds.), *Proceedings of the Joint Meeting of the 32nd Conference of the International Group for the Psychology of Mathematics Education, and the XX North American Chapter* (Vol. 2, pp. 423–430). Morelia, Mexico: University of Michoacan.

Ferrara, F. (2006). Remembering and imagining: Moving back and forth between motion and its representation. In J. Novotná, H. Moraová, M. Krátká, & N. Stehlíková (Eds.), *Proceedings of the 30th Conference of the International Group for the Psychology of Mathematics Education* (Vol. 3, pp. 65–72). Prague, Czech Republic: Charles University, Faculty of Education.

Gallese, V., & Lakoff, G. (2005). The brain's concepts: The role of the sensory-motor system in conceptual knowledge. *Cognitive Neuroscience, 22,* 455–479.

Goldin-Meadow, S. (2003). *Hearing gesture: How our hands help us think.* Cambridge, MA: Harvard University Press.

Kendon, A. (2004). *Gesture: Visible action as utterance.* Cambridge, England: Cambridge University Press.

Kress, G. (2000). Multimodality: Challenges to thinking about language. *TESOL Quarterly, 34*(2), 337–340.

Kress, G. (2001). *Multimodal discourse: The modes and media of contemporary communication.* London, England: Arnold Publishers.

Kress, G. (2005). Gains and losses: New forms of texts, knowledge, and learning. *Computers and Composition, 22,* 5–22.

Kress, G., Charalampos, T., Jewitt, C., & Ogborn, J. (2006). *Multimodal teaching and learning: The rhetorics of science classrooms.* New York, NY: Continuum International Publishing.

Lakoff, G., & Núñez, R. (2000). *Where mathematics comes from: How the embodied mind brings mathematics into being.* New York, NY: Basic Books.

McNeill, D. (1992). *Hand and mind: What gestures reveal about thought.* Chicago, IL: Chicago University Press.

McNeill, D. (2005). *Gesture and thought.* Chicago, IL: Chicago University Press.

Nemirovsky, R. (2003). Three conjectures concerning the relationship between body activity and understanding mathematics. In N. A. Pateman, B. J. Dougherty, & J. T. Zilliox (Eds.), *Proceedings of 27th Conference of the International Group for the Psychology of Mathematics Education* (Vol. I, pp. 105–109). Honolulu, HI: University of Hawaii.

Nemirovsky, R., & Ferrara, F. (2009). Mathematical imagination and embodied cognition. *Educational Studies in Mathematics, 70*(2), 159–174.

Radford, L. (2008). A cultural theory of learning. In L. Radford, G. Schubring, & F. Seeger (Eds.), *Semiotics in mathematics education* (pp. 215–234). Rotterdam, the Netherlands: Sense Publishers.

Radford, L. (2009). Why do gestures matter? Sensuous cognition and the palpability of mathematical meanings. *Educational Studies in Mathematics, 70*(3), 111–126.

Radford, L., Edwards, L. D., & Arzarello, F. (2009). Introduction: Beyond words. *Educational Studies in Mathematics, 70*(2), 91–95.

Robutti, O. (2006). Motion, technology, gesture in interpreting graphs. *The International Journal for Technology in Mathematics Education, 13*(3), 117–126.

CHAPTER 6

THE GESTURES OF BLIND MATHEMATICS LEARNERS

Lulu Healy
Solange Hassan Ahmad Ali Fernandes
Universidade Bandeirante Anhanguera, Brazil

When we began our investigations of the practices of blind mathematics learners in 2002, we had not envisaged that attending to their gestures would come to assume such a central role in our attempts to understand the particularities of their mathematics practices. Yet in retrospect, perhaps it could hardly have been otherwise; in the absence of access to the visual field, the hands and other parts of the body become critical channels through which information about the world is sensed, processed, and conceived. Still, although we know that blind learners will use their hands as substitutes for their eyes in order to perceive spatial data, does it necessarily follow that focusing on their gestures will illuminate the ways by which they make sense of mathematics? In this chapter, we will argue that it does. To do so, we first need to consider what we include as gestures.

Emerging Perspectives on Gesture and Embodiment in Mathematics, pages 125–150

WHAT COUNTS AS GESTURE?

In the mathematics education literature, the work of McNeill (1992) has had a major influence on those who have sought to investigate gesture use in the context of doing and learning mathematics (e.g., the special issue of *Educational Studies in Mathematics*; Radford, Edwards, & Arzarello, 2009). McNeill defines gestures as "movements of the arms and hands... closely synchronized with the flow of speech" (McNeill, 1992, p. 11). This definition stresses the speech-gesture relationships, and, as Roth (2001) suggests, many of those interested in gesture research limit themselves to "the study of hand and arm movements that are interpreted by others as part of what a person says" (p. 368). In McNeill's (1992) definition, then, gestures are to be treated as part of the same semiotic system as speech or, more precisely, as "elements of a single integrated process of utterance formation in which there is a synthesis of opposite modes of thought—global-synthetic and instantaneous imagery with linear segmented temporally extended verbalization" (p. 35). But if we are to see gesture as the visual manifestation of the imagistic aspects of cognition, then can we expect gestures to be used in the same ways by the blind and the sighted?

Indeed we might ask how global-synthetic, instantaneous imagery comes about for those who see with their hands rather than eyes because at least some aspects of spatial data are experienced in different ways when hands are used to "see." Both the blind and the sighted use what we might term "tools for seeing" to scan or move over physical objects or raised representations, in an intentional manner, taking in particularities of the forms in order to perceive, and at the same time conceive, the object (Ochaita & Rosa, 1995). A difference when hands are involved instead of eyes is that this process occurs in a slower and more successive manner. Whereas vision is synthetic and global, touch permits a gradual analysis from parts to the whole. Given the fact that perceiving through touch involves a sequential aspect that McNeill associates with the verbal, we wonder whether separating imagistic and verbal aspects of cognition makes sense in the case of the blind learner. This said, the way that information is conveyed in gesture does feel different from how it is communicated verbally. Gestures are dynamic expressions in which spatial and temporal relations can be preserved and communicated simultaneously. These features hold whether a gesture is experienced visually or through touch.

This brings us to another point. Not all body movements that we might call gestures—and certainly not all the hand movements that we have observed blind learners to make—occur in relation to speech. While still limiting attention to specific parts of the body, Alibali (2005) offers a more encompassing definition in which "gestures are defined as movements of the hands and arms that are produced when engaged in effortful cognitive activity" (p. 309). Using this definition, we might include as gestures the ways physical objects are explored by the hands, something we have chosen to do in this chapter. Roth and Thom (2009) also characterized as gestures the hand movements made by a young sighted learner as she attempted to characterize and classify three-dimensional geometrical objects.

However, even if we choose not to look at gesture exclusively as an accompaniment to speech, we still might not wish to include all hand and body movements as gestures. Rather, we concur with Rotman (2009), who calls a gesture "any body-movement that can be identified, repeated, and assigned significance or affect as a sign, a function, or an experience" (para. 3). In this definition, he divides gestures along three modes of embodiment. He associates the first, the semiotic mode, with forms of conveying meaning, significance, and affect. In the second, the instrumental mode, the body becomes akin to a machine. His examples include "pounding a nail" and "slicing an apple." The third mode, which he describes as immersive, refers to the experiencing/participating body and activities such as beating out rhythms, marching, and smelling flowers. In this chapter, we will center mainly on the semiotic body, but we will also consider how gestures emerge in relation to the body as instrumental and immersed in cultural activities.

Speech as Gesture

In relation to gestures as signs, we have also been influenced by Rotman's position. As he considers speech-gesture relations, he argues against treating gestures as some kind of semiotic supplement to speech, offering a view that differs from the combination of image and word suggested by McNeill. For Rotman, speech is species of gesture, perceived by auditory rather than visual means, but a gesture nevertheless. Once again, his point of view brings up questions for us because for those who have always been blind, gestures will never have been experienced visually. Nonetheless, we find his position attractive in that, by emphasizing gesture as "a body thing," he stresses the embodied nature of our sense-making experiences within the world. Indeed, Rotman is not the first to subordinate speech to gesture. The phenomenologist Merleau-Ponty (1945/1962) made a similar claim: "[T]he spoken word is a gesture, and its meaning, a world" (p. 215). Moreover, for him, the importance is not so much the form in which a sign representing experience is expressed, it is what this sign incites, that is, how it is felt by the interlocutors:

> I do not see anger or a threatening attitude as a psychic fact hidden behind the gesture, I read anger in it. The gesture does not make me think of anger, it is the anger itself. [...] The sense of gestures is not given, it is understood,

> that is, recaptured in an act on the spectator's part. [...] The communication or comprehension of gestures comes about through the reciprocity of my intentions and the gestures of others, of my gestures and intentions discernible in the conduct of other people. It is as if the other person's intention inhabited my body and mine his. (pp. 214–215)

If we go even further back in time to the Age of Enlightenment, we find a view of gestures as signs that originate in social practices. LeBaron and Streeck (2000) recount how in the 18th century, the sensualist Condillac described signs as "transformed sensations," and they describe how, for him, this transformation of embodied experiences into shared material signs holds the key to human knowledge:

> In his [Condillac's] view... [t]he formation of a symbol is a defining moment in the fabrication of shared knowledge because it allows the participants to focus upon and re-invoke previously shared experiences and to plan and conduct shared activities in their wake. (pp. 118–119)

According to Condillac, then, the language of action—that is, gesture—constituted the original, natural language. Here again, gesture is not reduced to an appendage of speech, instead it is seen as what gave rise to it.

We end this brief travel back in time by observing that, in the positions of both the sensualist Condillac and the phenomenologist Merleau-Ponty, we find early expressions of the thesis that characterizes what today we call embodied cognition—the idea that even the most abstract of symbols have bodily grounding. If such a claim might have been difficult to explore in previous centuries, this is no longer the case, with a growing body of evidence to support the idea that all human cognition is embodied (Anderson, 2003; Clark, 1997; Lakoff & Johnson, 1999; Varela, Thompson, & Rosch, 1991). The way we think cannot be separated from the way we act, and both have their bases in our human body, its sensory and motor capacities, and its location in space and time. Mathematics education too has entered the new era of the body, with an ever-increasing presence of embodied approaches (Lakoff & Núñez, 2000; Nemirovsky & Borba, 2004; Radford, 2009; Roth, 2010; Roth & Thom, 2009). Yet given the importance of symbol systems in accessing, or perhaps better in molding, mathematical activity, a characteristic of research in mathematics education that draws on embodied cognition is a tendency to attempt to develop approaches that also take into account the centrality of mediation (e.g., Arzarello, Paola, Robutti, & Sabena, 2009; Radford, Bardini, Sabena, Diallo, & Simbagoye, 2005). The attempt to combine embodied cognition with semiotic (and indeed material) mediation characterizes the approach we are adopting in our work with students who lack access to one or another perceptual field. We have become increasingly convinced by the argument that our mathematical understandings are structured by our encounters and interactions with the worlds we experience via our bodies. What we come to express symbolically results from transformations of our perceived experiences into sharable signs. The shareable nature of these signs reminds us also that mathematics as a discipline is a cultural affair; learning involves a delicate process in which learners become increasingly aware of how their subjective understanding for a given mathematical object connects to the cultural meanings emphasized in teaching.

MULTIMODAL IMAGININGS

In relation to embodied cognition, our view is that mathematical understanding can be expected to be a multimodal affair. Gallese and Lakoff (2005) offer neurological evidence to support this position, arguing that circuitry across brain regions links different sensorial modalities, "infusing each with the properties of others" (p. 456), and that both doing something and imagining doing the same something involve a shared neural substrate. This latter point is particularly relevant in the case of mathematics learners, where imagining is frequently revered as the means of engaging with the abstract, whereas doing is often seen as a less sophisticated interaction with concrete situations. In contrast to approaches that imply that mathematics learning is a process of decontextualization and in which abstraction is conceptualized as detaching objects, properties, and relationships from the contexts in which they were encountered, we see coming to mathematical understandings rather as involving a synthesis of concretion and abstraction. More specifically, we argue,

> concretion is posited as the making of personal connections, whereby learners express their own sense of an experience in a specific mathematical situation by relating this experience to some other (Hoyles & Noss, 1992; Wilensky, 1991). Abstraction, in its turn, is regarded as some conscious appreciation by learners of the generalised relationships implied in their expressions (Mason, 1989; Noss & Hoyles 1996). (Healy & Fernandes, 2011, pp. 160–161)

Concretion and abstraction, then, represent two faces of the same process. Concretion cannot occur without the abstraction of what is particular to the experiences, and abstraction is always contextualized. Learners continue to sense, to feel, the multimodal experiences involved in the process of identifying generalities. In this sense, we believe that thinking is also feeling. Indeed, we see gestures as one source of evidence that thinking involves imagined re-enactments of previously experienced activities. Where gestures accompany the noticing of generalized mathematical relationships, we have argued that they serve as embodied abstractions, acting as physical traces of the re-enacted experiences that convey the sense, the feeling, of the relationships in question (Healy & Fernandes, 2011). Like Radford (2009), we don't believe that gestures should be seen as "'windows' on inner thoughts or as conveyors of ideas that are already somewhere in the mind awaiting the proper material, namely, verbal expression" (p. 113) but as thinking-in-action.

This leads us also to question the nature of concepts, mathematical or otherwise. For us, they cannot be mental representations in which the abstract, logical universal properties of an object are stored in a somehow transcendental form stripped of the particularities of the settings in which it was encountered. Rosch (1999) presents a view that we see as an interesting alternative. For her, "concepts and categories do not represent the world in the mind; they are a participating part of the mind-world whole" (p. 70). Concepts in this perspective resemble what Barsalou (2009) defines as simulators, that is, distributed neural systems in which the multimodal content associated with instances of a category are integrated. These systems enable us to interpret new instances as tokens of a type (Barsalou, 2003). In this light, understanding anything can be defined as activating and reliving—that is simulating—any part of the previous experiences that have come to be allied with it. Moreover, this definition is just as applicable when what we are trying to understand is something immediately perceivable by our sensory apparatus as it is to something like a mathematical object, which manifests itself only through semiotic sign systems.

In other words, when we come into contact with any object (material or virtual), we reactivate a plurality of our perceptions of it, or, as Roth and Thom (2009) have suggested, we experience a single instance as a whole. This process of reactivation occurs on the basis of intentional actions, whose purpose is to recognize or cognize the object and from which are created a web of relations among the subject (body and mind), the object, and the world.

Here we might pause to consider someone who is not born blind but who loses the ability to see after a myriad of visual experiences. Our conjecture is that even though this individual is blind, she or he will still express a sense of experiencing things visually. Indeed, we have some supporting evidence for this conjecture in the reflections of a young man, Edson, who lost his sight completely at the age of 15. He explained to us that not only did he have visual memories, but also that he used them to recognize shapes, now experienced through touch, which he had learned about when he could still see. When we observed his interactions with symmetrical figures, this description did indeed seem to characterize his activities. During these interactions, as he moved his hands intentionally over a geometrical figure, represented by segments of elastic on a pin board, he told us how this allowed him to imagine the figure visually and hence operate on it "in his head." At the moment of exploration, Edson's gestures enabled him to recognize a shape as an instance of something he had previously seen (this example is presented in more detail in Fernandes & Healy, 2010a).

In short, we are arguing that our perceptive experiences enable us to form networks of multimodal neural systems that are stored in our memories. These are subsequently reactivated to simulate the perceptive, motor, and introspective states associated with an object even when it is not present. That is, the states associated with past experiences of a category of knowledge are relived through simulation (Barsalou, 2008). At times, gestures appear as spontaneously made physical traces of this process. Our interpretation is that, through these simulations, an individual is able to make present, through a temporal activation, cerebral processes created during past experiences.

The idea of temporal activation is not limited to those interested in embodied cognition. We also find it in Vygotsky's (1930/1978) ideas about the developing child, although characteristically in his version, spoken language is attributed prize of place:

> the child, with the help of speech, creates a time field that is just as perceptible and real to him as the visual one. The speaking child has the ability to direct his attention in a dynamic way. He can view changes in his immediate situation from the point of view of activities, and he can act in the present from the viewpoint of the future. (pp. 35–36)

What Vygotsky appears to be arguing here is that by using semiotic resources, the child becomes able to act in the present, drawing on the experience of past activities while attending to an endpoint, or goal, for the future. In this particular quote, he emphasizes the mediating role of speech in enabling the learner to use what she or he already knows (i.e., results of the past) to realize in the present (i.e., results of today) an activity that will come to characterize her or his own future performances (i.e., results of tomorrow).

Sense Organs as Mediating Tools

The idea of mediation is not limited to language use or even to semiotic and material tools. In fact, we have been struck how, in his early formulations of the notion of mediation, which occurred as he was working with learners with disabilities, Vygotsky began to discuss the idea that organs of the body, such as the eye, the ear, the skin, and the hands, can be thought of as "instruments" used to sense the world. In the absence of one functioning sensory apparatus, another instrument might be sought as substitute (Vygotsky, 1997). In this way, the hands of the blind might be considered as substitutes for the eyes. Hence, Vygotsky's writings suggest that he attributed the role of psychological tool, at least implicitly, to elements of the body.

Here, too, we identify a strong allusion to the embodied nature of the human intellect, but it is important to remember that Vygostky was primarily interested in the sociocultural nature of tools, created as they are in response to the demands of particular practices, which characterize particular moments in the history of a given culture. Indeed, for Vygotsky, a defining human quality is the need and ability to both use artifacts to mediate activities and encourage the appropriation of these forms of mediation by subsequent generations (Cole & Wertsch, 1996). For him, such artifacts both transform the activities of those using them and can be transformed in use. Because organs of the body can be thought of as mediating tools, the substitution of the eye by, say, the hand can be expected "to cause a profound restructuration" (Vygotsky, 1997, p. 99) of the intellect and of the personality of the blind individual. This means that to understand the practices of blind learners, we cannot leave to the side the (multi) sensorial forms by which they have access to learning artifacts. Our thesis here is that the sensorial apparatus impacts on learners' understandings as much as do the other material and semiotic artifacts that compose the experienced world.

As we attempt to adopt a sociocultural stance in our investigations of the mathematics learning processes of blind learners, we recognize that cognition is not only situated historically and culturally but also bodily. This means that we need to attend to the perceptual activities, the signs, and the materials employed by the blind learners as they express their developing understandings of the content under study. Hence, gestures become a central focus in our analyses.

There is currently rather little research into gesture use by blind individuals. The studies we have located do not relate directly to gesture use in mathematical activity, but the evidence they report suggests that gestures accompany the communicative practices of the blind as well as the sighted (Goldin-Meadow, 2003; Iverson & Goldin-Meadow, 1998). The research of Iverson and Goldin-Meadow (1998), for example, investigated the frequency and type of spontaneous gestures that occurred in dialogues involving only the blind, only the sighted, and both blind and sighted individuals. They found no significant differences among the three groups in the number of gestures used during the dialogues, although they did observe that the gestures of the blind tended to be limited to deictic gestures (i.e., gestures that indicate), beat gestures (i.e., gestures that accompany the rhythm of the discourse), and iconic gestures (i.e., gestures that represent concrete objects and actions). On the basis of these results, Goldin-Meadow (2003) conjectured that what McNeill (1992) has classified as metaphorical gestures, or gestures that represent abstract ideas pictorially, tend not to be produced by blind individuals. It may be, however, that these results were a function of the contexts investigated and the content of the dialogues involved. In our work with blind adolescents carrying out a number of mathematical tasks, we have observed gestures to play as much a part in communications about abstract entities and relationships as in concrete ones (Healy & Fernandes, 2011). Taken together, these results suggest that, even among those who were born blind and who have never visually seen the gestures of others, communicative hand and arm movements accompany their speech, and the kinds of gestures they produce are contingent on the context of their interactions. In fact, because of the ways these students use their hands as substitutes for their eyes, their gestures may reveal even more about their intentions as they explore mathematical objects and the nature of the simulations through which they seek to make sense of these objects than do the gestures of the sighted.

GESTURE USE BY BLIND MATHEMATICS LEARNERS: EXPLORING AREA AND VOLUME

In the rest of the chapter, we turn our attention to two episodes from a sequence of learning situations undertaken with a group of four learners who attended a regular school, which was part of the public education system of the State of São Paulo in Brazil. The learners, aged between 14 and 18 years, were first-year high school students. Marcos (18 years old), Fabio (16 years old), and Caio (17 years old) were all born with different kinds of congenital blindness, whereas Leandro (14 years old) had lost his sight at the age of two in an accident that resulted in the severing of his optic nerve. The data we report are drawn from a sequence of activities associated with the study of volume, area, and perimeter. The activities were designed to include tools intended to favor tactile exploration of the mathematical objects in question. The ways in which the tasks were proposed aimed to stimulate dialogues among the participants; the examples here focus on the interactions between researchers, acting also as teachers, and the blind learners. The sequence of activities was implemented over four research sessions, each approximately 90 minutes in duration. The activities took place in a room that houses the resources specifically designed for the blind students and substituted their regular mathematics lessons. All the research sessions were videotaped and transcribed for analysis.

The activities were designed to involve students in distinguishing among perimeter, area, and volume and in developing methods to calculate them. More specifically, in relation to area, the tasks had been designed to integrate three interrelated aspects that Kordaki (2003) argues are essential to develop understanding of area: the conservation of area, area measure, and formulae for calculating area. The idea was that the strategies developed for area might be generalized to volume.

The students had explained to us that it was rare for them to interact with representations of geometrical shapes, and an important aspect of designing the tasks was to produce tactile materials that would make this possible. During the first session, they were given a board with the impressions of four rectangles, which could be filled with either wooden unit cubes or rectangular and triangular shapes in foam rubber. In this session, the students explored the areas of rectangles and triangles. In particular, they experienced how the area of a triangle could be perceived as half the area of a rectangle with the same height and base (for a more detailed analysis of the activities during this session, see Fernandes & Healy, 2010b).

The second session was dedicated to determining the areas of the plane figures represented in foldable cardboard shown in Figure 6.1 (with the numbers indicating the order in which they were presented to the students). During this session, the students worked with adapted rulers in which the number marks were raised so they could be read tactilely.

In the third session, the students worked on determining the surface areas and volumes of a range of square and rectangular boxes (initially presented as two-dimensional cardboard nets). Once again, unit cubes were on hand as a possible means of measure, as were the adapted rulers. In the fourth session, students worked again with nets, which this time could be folded into the form of oblique prisms.

The two episodes analyzed in this chapter have been chosen to illustrate different aspects of gesture use that emerged during these activities. To aid in interpreting the transcriptions that follow, Figure 6.2 shows how the students and researchers were arranged around a table as the first episode occurred.

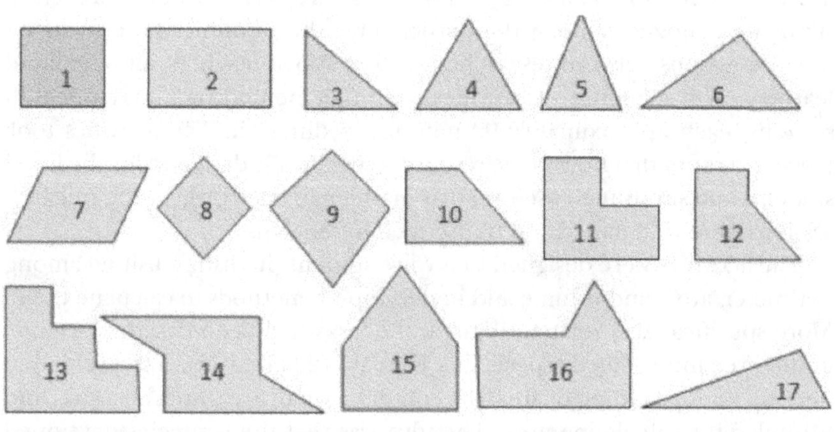

Figure 6.1 Cardboard shapes used in the second research session.

Figure 6.2 The research participants in clockwise order, starting from bottom left: Marcos, Leandro, Researcher 1 (Lulu), Researcher 2 (Solange), Caio, Fabio.

Example 1: Gesturing for Oneself

This episode occurred during the second session, when the students were working on finding the areas of plane figures. More precisely, it centers on the interactions during the students' attempts to determine the area of the seventh shape shown in Figure 6.1: a parallelogram with a length of 12 cm and a height of 7 cm. Each student received his own cardboard representation. Perhaps because the four previous shapes they had worked with were triangles, the first suggested area was 42 cm². This area had been obtained by measuring the length of the parallelogram (which they called the "base"), its height (correctly located by measuring the perpendicular distance from the base to the vertex on the opposite side), and then dividing by 2. Just as Caio was confirming to the researchers that this was the area on which they were all agreed, Fabio, who had been one of the initiators of the "dividing by 2" strategy, began to have doubts.

As he said this, Fabio was holding the parallelogram in two hands, such that both of his thumbs rested along the base, and his index fingers pointed upward as if they were perpendicular to the base rather than following the edges of the parallelogram (Figure 6.3). That is, the thumb and finger of Fabio's left hand made an L shape while the corresponding digits on the right hand made a reflected L shape. Our interpretation is that this gesture represents the visible trace of an imagined action in which the sides of the parallelogram are "straightened" to create a rectangle. This was not immediately obvious; his intention was revealed only gradually to us as the dialogue continued. We believe that Fabio, at this point, made the gesture

Caio: Area, 42
Fabio: Hey, you know, I think I am being a bit of a donkey.

Figure 6.3 Fabio thinks about rectangles.

for himself and that there was no conscious intent on his part that the gesture would serve as part of the way of communicating his developing idea to others. It is part of his thinking-in-action.

As Caio continued to explain the first idea about area, Fabio pondered a moment longer before quietly offering a new strategy, which emerged along with his gesture.

Challenged to explain his idea, Fabio was a bit more explicit about his intentions. His tone suggested that he was not entirely convinced by the idea, but as he moved his hands to express an almost identical gesture again (Figures 6.3 and 6.4 show essentially the same gesture of an L shape and its reflection), he seemed to gain confidence.

Caio: 7 times 12, the base times the height. . . 7 times 12, 84, divided by 2, 42.
Fabio: We could do, we could do 8 times 12
Marcos: Why? Why 8?
Leandro: Why 8?
Fabio: Because if we straightened it up, it would be a rectangle. [makes the gesture in Figure 6.4]

Figure 6.4 Fabio's gesture reappears.

Fabio: Yes, it would be a better-behaved area.
Caio: A what?
Fabio: A better-behaved area.
Caio: Ah I get it.

By this point, Fabio seemed surer of his strategy, and although his gestures remain unseen by the other blind students, his mention of a rectangle and a "better-behaved" area also provoked a change in Caio, who no longer thought they should divide by 2 but continued to be worried about using a measure of 8 rather then 7.

Fabio: 12 times 8
Caio: 12 times 8, so if it was 12 times 7?
Res 2: Why 12 times 7?
Caio: Because it's this base [runs his finger along one of the longer sides of the parallelogram] times

Caio began to explain to the researcher, using deictic gestures to indicate the side he was using as the base. He did not just point to the side but ran his finger along the entire segment in a dynamic form that was characteristic of the way participants (both the blind students and sighted researchers) indicated geometrical objects during this session. Caio never finished his explanation; he was interrupted by Leandro, who wondered why they were no longer dividing by 2. At this point, as three possible areas were still being contemplated (42cm^2, 84 cm^2, and 96 cm^2), Fabio was questioned about his strategy again.

This last exchange leaves it clear that Fabio was not confusing the side of the parallelogram with its height. He knew that the height of the parallelogram—as it was—was 7cm. However, as his gesture consistently implied, he had another shape in mind: the rectangle that would result from a rotation of the sides, of length 8, until they form an angle of 90° to the sides of width 12. The gesture appears then to express, physically, a generalized relationship by which a parallelogram might be transformed into a rectangle—that is, it expresses an embodied abstraction. The consistent use of the same gesture suggests it had become a sign that represents for Fabio a kind of area that is easy to measure.

This sign later became associated with a verbal one, as Fabio began to talk of a "better-behaved" area. Moreover, as Fabio's abstraction was articulated verbally, he re-enacted once again the gesture, stressing how he continued to physically sense the object he was describing verbally and, we would argue, showing the connection between linguistic and kinesthetic expressions. Regardless of the different forms of expression, verbal or gestural, the abstraction was embodied. There was still a problem, though; Fabio's

Res 1: I don't understand why 12 by 8.
Fabio: It's because, if it was straight, it would be a rectangle [makes the gesture of the L-shape and its reflection in Figure 6.5]. And the area of a rectangle is base times height, isn't it?
Res 1: Uh huh, and what is the height and what is the base?
Fabio: The height would be 8... if it was straight.
Res 2: But since it isn't, what is the height?
Fabio: 7.
Caio: So. So, why are you, why don't you [laughs]
Fabio: Aiee [laughs]

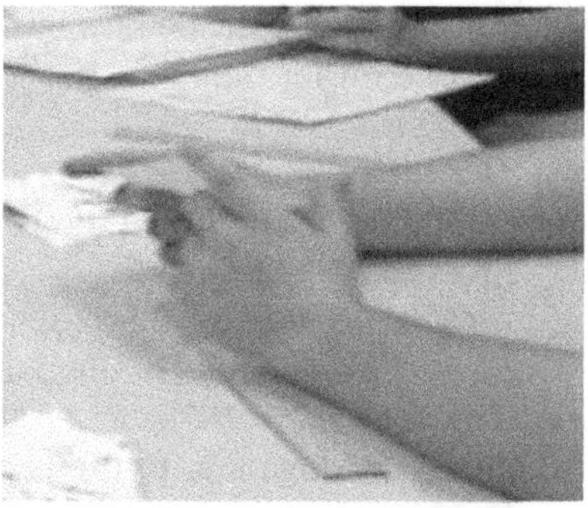

Figure 6.5 The gesture of the L shape and its reflection appears for a third time.

idea to transform the parallelogram into a shape whose area he knew how to determine was a sensible one, but he did not seem to realize that his transformation would have the result of changing the measures in question.

An impasse seemed to have been reached. Perturbed perhaps by his own reflections about the two possible heights of the figure he was thinking about, Fabio suggested jokingly they go back to the value of 42 for the area of the parallelogram. Marcos found this attractive because the shape certainly has what he described as "triangle parts"—as he considered the value 42, he ran his thumb and forefinger simultaneously toward one of the corners of his parallelogram (Figure 6.6a). Leandro also used a similar gesture in his explorations (Figure 6.6b). Perhaps this reliving of a sensation that had previously accompanied their explorations of triangles (in Barsalou's terms, a reactivating of a simulator associated with their multimodal experiences of triangles) reinforced the selection of the inappropriate method.

As all four students went back to tactile explorations of the parallelogram, Caio and Fabio became convinced that dividing by two was no longer an appropriate option. Caio stated, "It seems like a triangle, but it isn't a triangle, it's something close to a rhombus or something like that." Fabio stated, "We are not working with a triangle anymore; we are working with a whole figure." For Caio, like Marcos and Leandro, there were similarities in

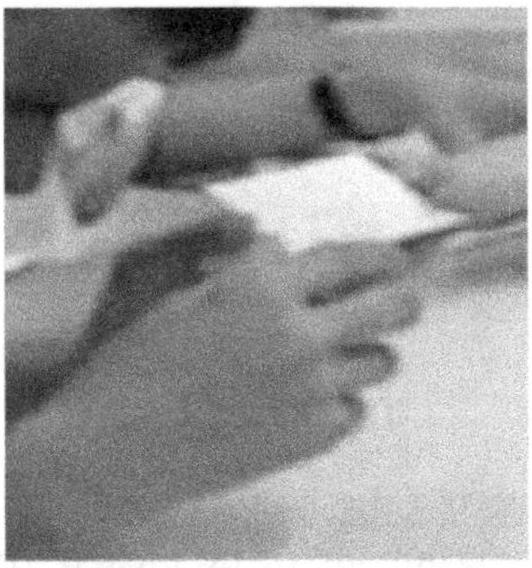

Figure 6.6a Marcos feeling with his thumb and forefinger a triangular-like corner.

Figure 6.6b Leandro using an equivalent gesture to feel a corner of his parallelogram.

the shape they were perceiving and what they had perceived when exploring triangles, but we find Fabio's comment particularly revealing. Again, his words suggest the simulation of a previous experience, an experience all four students had shared in the prior research session. During this session, triangles were perhaps not perceived as "whole figures" because, in the context of the area activities, their area had been introduced as half of a rectangle. By choosing to describe the shape in this way, Fabio brought to the present aspects of a previous activity, in which, he knew, all the other participants had shared. Caio went on to use the same combination of words to once again offer his area calculation "12 times 7, which is base times height, and since it's a whole figure you don't divide by 2." However, although the group did resolve the issue of whether to divide by 2, there seemed no easy way to decide whether the base should be multiplied by 8 or by 7, and at this point, Researcher 1 decided to intervene:

> **Res 1:** I'm going to do something that involves destroying Leandro's figure.
> **Leandro:** Destroying my figure? [laughs]
> **Res 1:** Can I? I'm going to cut it.

With Leandro's permission, Researcher 1 decomposed the parallelogram into a right triangle and a trapezoid (Figure 6.7).

She then placed the two pieces into Leandro's hand, tracing his index figure along side of the trapezium congruent to the diagonal of the right triangle and then along the diagonal in turn, as she suggested he fitted the two pieces together so that these two segments coincided. Here, we note another characteristic of gesture use during interactions with the blind.

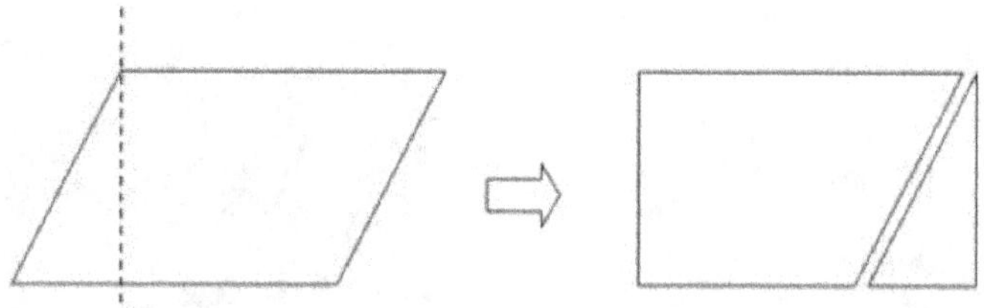

Figure 6.7 Decompostion of parallelogram.

When there was a truly conscious attempt to use the gesture as a means to communicate, the gesturer not only pointed, but she or he picked up the hand of the other and enacted or re-enacted the gesture in question with the intention of enabling the other to share both the idea and the feeling.

Res 1: Now this part here [traces Leandro's figure along the side of the trapezoid], and [traces his finger on the congruent side of the triangle as in Figure 6.8].

Leandro: Oh look! That makes more sense. It's great now. Great.

Figure 6.8 An intentionally communicative gesture.

While Researcher 1 was working with Leandro, Fabio re-convinced himself about the value of 96 for the area of the parallelogram, tapping the figure repeatedly on the table as he almost chanted "96, 96, 96, 96, 96, 96, 96." This action provoked Researcher 2 to pass him one of the rectangles he had previously worked with, which fortuitously also had a height of 7 cm. Fabio placed the rectangle on top of the parallelogram, as shown in Figure 6.9, and then moved his hand briefly over the triangle shapes that remained from the original parallelogram (Figure 6.10).

He then went on to measure the height of the rectangle before handing it back to Researcher 2, saying, "I've changed my opinion. The area is 84. 12 times 7. The height is 7." Although we cannot be completely sure, the introduction of the rectangle appeared to offer Fabio a new means by which he might transform the parallelogram into a shape with a "better-behaved" area. In fact, after his change of mind, Researcher 1 passed him the two pieces she and Leandro had been working with. Fabio confirmed, although

Figure 6.9 The partially covered parallelogram.

Figure 6.10 Fabio feels the uncovered triangle.

with rather less enthusiasm than Leandro, that they "make a rectangle, like I said" and handed the pieces back to the researcher.

As we reflect on the interactions that occurred during this episode, we note that, in general, there was little evidence of any reciprocity in the gestures of the students. With the exception perhaps of the exchange between Researcher 1 and Leandro, it did not seem that the intentions of one participant came to inhabit the body of another, as Merleau-Ponty describes. But this does not mean that gestures did not play any role in the mathematical practices that emerged. On the contrary, Fabio's gestures leave a trace, to those of us who can see them, of his attempts to transform the parallelogram into a shape whose area he already knew. Fabio's gestures seem to have been directed at himself, helping him feel a transformation as he imagined it. Indeed given the tools he had in hand, because the transformation he had in mind was not physically possible, imagining was the only option.

Gestures were also central in the students' attempts to recognize the figure. In this case, they were not always efficient, as feeling the pointed corner of the parallelogram conjured up sensations associated with triangles. Another notable point is that even though the students could not see each other's gestures, there were certainly times in which similar gestures emerged. Perhaps this is only to be expected given the way our hands work, but we were surprised by the similarities in the ways the sighted researchers and the blind students moved their hands over the geometrical shapes. For example, rather than pointing to a static point on a side (even if the side was being indicated only to oneself), the fingers were moved dynamically, tracing along the entire side. When vertices were explored, this too was done in a dynamic manner, running two fingers simultaneously to the vertex rather than pointing in a discrete manner to it. Of course, for blind learners, continuous gestures make sense because it would be harder for them to point discretely to the end points of a segment than to trace the trajectory between them. Indeed, the very way that they use their hands as eyes confirms that they were using a way of perceiving that differs from the more synthetic visual view. But it is interesting that as researchers, we also did this—even when we were gesturing for ourselves. Perhaps these gestures are simply common for everyone, blind or sighted, or perhaps as we observed the students working, we had begun to appropriate for ourselves the ways that they perceived the objects. Perhaps our gestures were reflections of our attempts to feel their intentions.

This first episode serves to illustrate that even when there were few conscious attempts to communicate by means of gesture, gestures nonetheless emerged as part of thinking—that is, as traces of the imagined doings that were enacted during the search for solutions to the task in hand. It also shows how, when these gestures cannot be seen by the other participants, it can be hard for them to understand each others' intentions. The second episode illustrates a rather different case.

Example 2: Gestures as Means to Feel Another's Feeling

This episode comes from the last of the four sessions. The group had already developed ways to find the volumes of cubes and rectangle-based prisms. During the interactions around these tasks, they had again talked in terms of the wooden unit cubes as units of measure, although they never actually handled the cubes in their discussions. Their gestures and speech suggested that they were thinking of the volume of the cuboids as layers of these cubes. The task we will consider in detail involved determining the volume of an oblique prism. As in the previous task, each participant had received an identical cardboard shape, in this case, the net of the prism shown in Figure 6.12. In contrast to the first episode, where everyone was working individually before sharing their results, during this task, the students spontaneously split into two pairs. Here we focus only on Fabio and Leandro. Leandro's net had been folded and taped to form the solid, whereas Fabio's remained open. After a period in which the pair worked separately collecting different measures of the prism, Fabio suggested to Leandro that he had arrived at a value for the volume.

As Leandro asked Fabio to explain, he held out his hands toward Fabio (Figure 6.11), who took Leandro's right hand in his own and then reached for the prism with his left hand. We might therefore interpret Leandro's request for an explanation as a request to feel physically what Fabio has been feeling. In the exchange that follows, Fabio seems to respond to this

Fabio: I think that 385 little cubes will fit, but now I need to explain how.
Leandro: Explain to me.

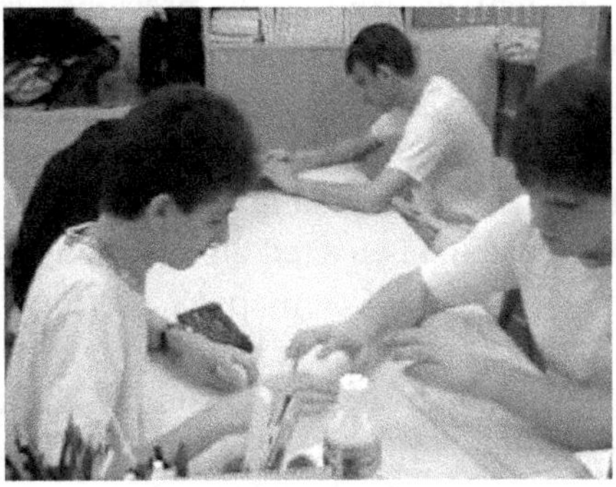

Figure 6.11 Leandro holds out his hand for an explanation.

request, as he moved Leandro's hand to reproduce the gestures he had made during his explorations.

> **Fabio:** [Placing Leandro's hand on one of the parallelogram faces of the prism] I took just the rectangle. I took off the triangles [as shown in Figure 6.13, Fabio traces Leandro's finger first up the line marked a in Figure 6.12 and them up the line marked b]. Then I did this here [Traces Leandro's index figure along the perimeter of the rectangle he has created], I found the area, it was 42, 7 times 6. Then I found the two triangles, 5, 5 . . .
> **Leandro:** And how much is the area?

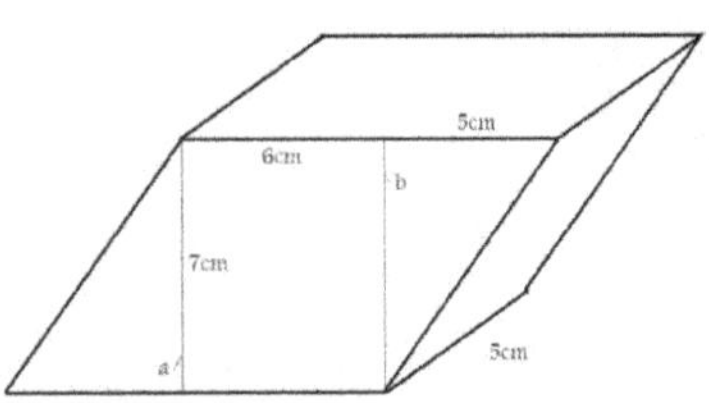

Figure 6.12 Fabio's decomposition of the prism.

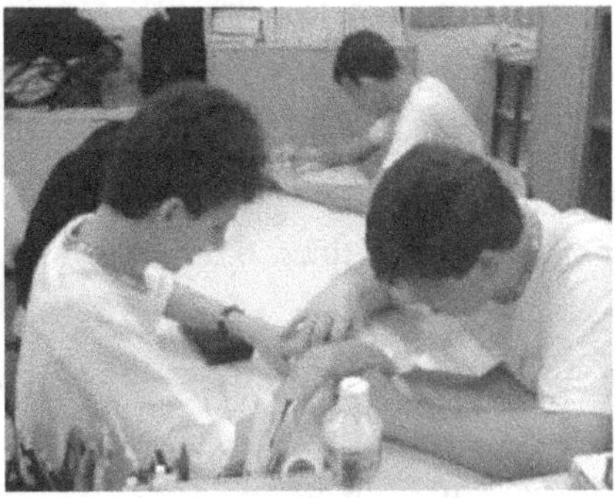

Figure 6.13 Fabio shows Leandro how the parallelogram was decomposed.

In marked contrast to previous activity, where Fabio's gestures were made principally for his own benefit, leaving his colleagues to guess how he perceived the object in his hand, in this episode, his gestures effectively communicated to Leandro the way he decomposed the three-dimensional shape. His gestures were aimed at helping Leandro feel the shape as he did, and as Fabio re-enacted his imaginings for Leandro, he was also re-enacting them for himself. Perhaps as a result of this bodily led explanation, Leandro did seem to follow Fabio's thoughts, and with the area of the rectangle determined, he wanted to confirm the area of each triangle. Fabio first suggested it was 25, as he confused the measure for height (7 cm) with depth (5 cm). As he checked, he took the prism from Leandro's hand and felt the height once again, keeping this checking gesture to himself (Figure 6.14).

A further point that we might make is that Fabio's decomposition of the prism during this episode appears to represent a generalization of what he had felt when he placed the rectangle on top of the parallelogram in the previous episode—he treats the parallelogram-shaped face of the prism as if it were composed of a rectangle and two triangles that fit together to produce another rectangle. Leandro had also previously experienced the cutting off of a triangle to rectangularize the parallelogram—and Fabio knew that Leandro had felt this—something else that might have made the sharing of the strategy more straightforward. In the end, Leandro finally announces that they have reached a conclusion. The task is not quite over, though, as the pair still needs to explain their strategy to the researchers. In

Fabio: 25. (Takes the prism back from Leandro) Wait a minute that's 7 (runs his finger along the height and corrects the previous calculation). 5 times 7, then times 5, 5 times 7 times 5

Leandro: 175.

Fabio: And 7 times 6 times 5, 210.

Leandro: 210.

Fabio: 385.

Leandro: [confirms using a speaking calculator] 210 and 175, 385. I think we have arrived at a conclusion.

Figure 6.14 Fabio checks the height for himself.

an interesting move, Fabio decided to use the net of the prism rather than the three-dimensional form when explaining to Researcher 1.

> **Res 1:** You cut...?
> **Fabio:** I did it like this. Look [picks up the net of the prism as shown in Figure 6.15]. I would fold it, in this case like this [folds along the height to form one of the triangles]. So, here I get the area of the rectangle. After that I calculated this part here [passes his hand over the triangle created by the fold, Figure 6.16], which is 5 by 5, I mean 5 by 7. So I have another imaginary triangle that would make a square of 7 by 5. That is, a rectangle.
> **Res 1:** So joining the two what you are calling triangles gives a rectangle...
> **Leandro:** 5 by 7
> **Fabio:** 5 by 7, and here [runs his figure along an edge representing depth] 5
> **Res 1:** And the volume this would be?
> **Fabio:** 175
> **Res 1:** And the other part?
> **Leandro:** 210
> **Res 1:** 210?
> **Fabio:** Because the rectangle is 6 by 7, which gives 42. Then I did the volume, times 5, is 210.
> **Res 1:** OK, I got it. I understand.

In this exchange, Fabio once again changed the style of his gestures. Because he knows that the researcher can see, it was not necessary to take her hands in his. We wonder why he decided to make use of the net in this

Figure 6.15 Fabio folds the net to show how he imagined the rectangle.

Figure 6.16 Fabio passes his hand over triangle so Researcher 1 can see it.

explanation rather than using the solid shape as he had done with Leandro. Was it perhaps because, while he wanted Leandro to feel the prism as he had felt it, his explanation to Researcher 1 had a different intention: He wanted her to know that he could see the prism using strategies she had previously helped him to feel? Whatever the reason for the switch, in this episode, we see how gestures can be used with an explicit communicative intent between the blind and the blind as well as the blind and the sighted. Moreover, in this case where Leandro could feel Fabio's physical explorations as well as hear the words that accompanied them, appropriating Fabio's thinking appeared to be much less arduous than in the previous episode where only verbal descriptions were offered.

REFLECTING ON GESTURE USE IN INTERACTIONS WITH BLIND MATHEMATIC LEARNERS

Our aim in this chapter has been to examine the role of gestures in the mathematical sense-making activities of learners who do not see with their eyes. By offering examples of gesture production in the mathematical practices of blind learners, we have attempted to characterize their learning processes in the light of a perspective that recognizes the situated and embodied nature of cognition. In particular, we hope that the examples presented in this chapter have shown that, as we seek to understand the process of learning of mathematics, we need to listen just as carefully to what learners say with their hands as we do to other forms of expression.

We have argued that as blind learners perceive mathematical objects, gestures serve as a means by which they can create and express embodied abstractions associated with the identification or definition of general properties and relationships. In particular, in both episodes presented in this chapter, gestures were fundamental in the abstraction of methods that permit the transformation of parallelograms into rectangles. We call these gestures abstractions because we believe they express relationships that are concrete in the domain in which they are made but which might be concretizable in other domains as well. Fabio's L-shaped gestures in the first episode embody his sense, his feeling, of the generalized relationships that he was noticing. Such gestures are not only material manifestations of aspects of physical activities, they also have a semiotic role, carrying a sense of the mathematical understanding that has come to be associated with the activities of those who make them.

In relation to the first episode, our view is that Fabio's gestures are essentially intended for himself, whereas in the second case he makes explicit attempts to enable Leandro to perceive them. In such a case, we can conclude they had an explicit communicational intent. Given the blindness of these students, it is rather easier to make this distinction. When all participants are sighted, even if the gestures of one are not specifically directed toward others, the others might see them and hence already begin to perceive what the gesturer perceives. This type of perception requires a different kind of activity—and the use of physical contact—in dialogues between the blind. Indeed, in the first episode, as sighted researchers, we had more sensory information about Fabio's thinking-in-action than the other participants, as we could see his gestures, even though they were not necessarily intended for us (and using the video recording we could also replay them many times over).

But independent of whether gestures are primarily intended for oneself or for others, the embodied abstractions they express can be thought of as social rather than individual acts—the imagined or actual actions involving the reactivation of perceptive, motor, and introspective states associated with such objects in previous activities that occurred in similar social settings. Such gestures can be seen as outward signs of imagined re-enactments of previous doings with things and with others—and hence as evidence of the cognitive similarities involved in doing something and imagining doing the same thing. The data presented in both episodes evinces the social nature of all the interactions with the signs created in bodily hand movements as well as in verbally formed words chosen so as to be shareable, even if they may end up being experienced most forcibly by the person who makes them.

One other observation from our analysis relates to the similarities in the gesturing activities of the blind students and the sighted researchers. As would be expected, when either student or researcher wished to explicitly communicate a feeling to one of the blind students, they picked up his hand in theirs and dynamically traced along the objects they were referring to; they did not just point to them. Perhaps a little less predictably, along the entire sessions from which the episodes were drawn, both researchers and students made gestures that they did not try to communicate explicitly to those who could not see or feel them. Most striking of all was that these gestures too were similar. The researchers consistently ran their hands along the edges they were talking about (from vertex to vertex), just as the blind students did, even when they were not purposely attempting to show the edges to others.

We offer two possible interpretations. Perhaps the ways we attempt to perceive things tactilely are independent of whether we can see them visually. Supporting evidence for this view can be found in the descriptions by Roth (2010) of a sighted learner's explorations of the relationships between rectangular prisms and cubes. Certainly, there do seem to be similarities between the gestures used by his sighted subjects and the blind students who have participated in our research. However, deictic gestures are usually more associated with pointing to a specific location of, or on, an object. Indeed, McNeill (1992) originally describes them thus: "deictic gestures are pointing movements, which are prototypically performed with the pointing finger"

(p. 80). Pointing is a little different from the dynamic tracing movements of the deictic gestures described in this chapter, which indicate not only the location of objects but also their location in relation to others. The first kind of "static" pointing gestures did occur in the settings we researched, but they were relatively rare. So a second interpretation is that the nature of the gestures produced by the blind students is influenced by the way they perceive—because they use their hands as substitutes for their eyes, a different kind of seeing is enabled, and the gestures evidence this difference. If this is the case, then perhaps the similarity in gestures results from the attempts of all the participants to share in the feelings of others, even though not all of us were experiencing the world in all the same ways. That is, in our interactions with blind learners, it was as important that we, as researchers who were also acting as teachers, tried to appropriate their ways of acting on the world as it was that we tried to enable them to appropriate ours.

Moreover, if, as Gallese and Lakoff (2005) claim, both doing something and imagining doing the same something involve activating the same multisensory cognitive resources, then it may be that as human beings, we possess bodily mechanisms that are tuned to sense the feelings of others, to feel almost as if we temporarily inhabit their bodies, as Merleau-Ponty (1945/1962) would have it. Hence, it is possible that we, the sighted researchers, came to use gestures that resembled those of our blind collaborators exactly because this was one of the means by which we could interpret (i.e., sense) what they were thinking. Perhaps one implication of our research then is that we should make more conscious efforts to ensure that the process of teaching is conducted in this empathetic manner that our bodies are wired to permit.

ACKNOWLEDGMENTS

We are grateful to the funding we have received from Fundação de Amparo à Pesquisa do Estado de São Paulo (Project No. 2004/15109-9) and Coordenação de Aperfeiçoamento de Pessoal de Nível Superior (Project No. 23038.019444/2009-33) in the course of this research program. We are also grateful to the students and teachers who so generously gave their time and energy to collaborate with us during the design and execution of the research activities.

REFERENCES

Alibali, M. W. (2005). Gesture in spatial cognition: Expressing, communicating, and thinking about spatial information. *Spatial Cognition and Computation*, *5*(4), 307–331.

Anderson, M. L. (2003). Embodied cognition: A field guide. *Artificial Intelligence,* 149(1), 91–103.

Arzarello, F., Paola, D., Robutti, O., & Sabena, C. (2009). Gestures as semiotic resources in the mathematics classroom. *Educational Studies in Mathematics, 70,* 97–109.

Barsalou, L. W. (2003). Situated simulation in the human conceptual system. *Language and Cognitive Processes, 18* (5/6), 513–562.

Barsalou, L. W. (2008). Grounded cognition. *Annual Review of Psychology, 59,* 617–645.

Barsalou, L. W. (2009). Simulation, situated conceptualization, and prediction. *Philosophical Transactions of Royal Society: Biological Sciences, 364,* 1281–1289.

Clark, A. (1997). *Being there: Putting brain, body and world together again.* Cambridge, MA: MIT Press.

Cole, M., & Wertsch, J. V. (1996). Beyond the individual-social antinomy in discussions of Piaget and Vygotsky. *Human Development, 39,* 250–256.

Fernandes, S. H. A. A., & Healy, L. (2010a). Embodied mathematics: Relationships between doing and imagining in the activities of a blind learner. In M. M. F. Pinto & T. F. Kawasaki (Eds.), *Proceedings of the 34rd Conference of the International Group for the Psychology of Mathematics Education* (Vol. 2, pp. 345–352). Belo Horizonte, Brazil: PME.

Fernandes, S. H. A. A., & Healy, L. (2010b). A inclusão de alunos cegos nas aulas de matemática: Explorando área, perímetro e volume através do tato. *Bolema: Boletim de Educação Matemática, Rio Claro (SP), 23*(37), 1111–1135.

Gallese, V., & Lakoff, G. (2005). The brain's concepts: The role of the sensory-motor system in conceptual knowledge. *Cognitive Neuropsychology, 22,* 455–479.

Goldin-Meadow, S. (2003). *Hearing gestures: How our hands help us think.* Cambridge, MA: Harvard University Press.

Healy, L., & Fernandes, S. H. A. A. (2008). The role of gestures in the mathematical practices of blind learners. In O. Figueras, J. L. Cortina, S. Alatorre, T. Rojano, & A. Sepúlveda (Eds.), *Proceedings of the Joint Meeting of PME 32 and PME-NA XXX* (Vol. 3, pp. 137–144). Morelia, México: Cinvestav-UMSNH.

Healy, L., & Fernandes, S. H. A. A. (2011). The role of gestures in the mathematical practices of those who do not see with their eyes. *Educational Studies in Mathematics, 77,* 157–174.

Iverson, J. M., & Goldin-Meadow, S. (1998). Why people gesture when they speak. *Nature, 396,* 228–228.

Kordaki, M. (2003). The effect of tools of a computer microworld on students' strategies regarding the concept of conservation of area. *Educational Studies in Mathematics, 52,* 177–209.

Lakoff, G., & Johnson M. (1999). *Philosophy in the flesh: The embodied mind and its challenge to western thought.* New York, NY: Basic Books.

Lakoff, G., & Núñez, R. (2000). *Where mathematics comes from: How the embodied mind brings mathematics into being.* New York, NY: Basic Books.

LeBaron, C., & Streeck, J. (2000). Gestures, knowledge, and the world. In D. McNeill (Ed.), *Language and gesture* (pp. 118–138) Cambridge, England: Cambridge University Press.

McNeill, D. (1992). *Hand and mind: What gestures reveal about thought.* Chicago, IL: University of Chicago Press.

Merleau-Ponty, M. (1962). *Phenomenology of perception* (Colin Smith, Trans.). London, England: Routledge and Kegan Paul. (Original work published 1945)

Nemirovsky, R., & Borba, M. (2004). Body activity and imagination in mathematics learning. *Educational Studies in Mathematics, 57*(3), 303–321.

Ochaita, E., & Rosa, A. (1995). Percepção, ação e conhecimento nas crianças cegas [Perception, action and knowledge of blind children]. In C. Coll, J. Palacios, & A. Marchesi (Eds.), *Desenvolvimento psicológico e educação: Necessidades educativas especiais e aprendizagem escolar* (Vol. 3, pp. 183–197) (Marcos A. G. Domingues, Trans.). Porto Alegre, Brazil: Artes Médicas.

Radford, L. (2009). Why do gestures matter? Sensuous cognition and the palpability of mathematical meanings. *Educational Studies in Mathematics, 70*(2), 111–126.

Radford, L., Bardini, C., Sabena, C., Diallo, P., & Simbagoye, A. (2005). On embodiment, artefacts, and signs: A semiotic-cultural perspective on mathematical thinking. In H. L. Chick & J. L. Vincent (Eds.), *Proceedings of the 29th Conference of the International Group for the Psychology of Mathematics Education* (Vol. 4, pp. 113–120). Melbourne, Australia: University of Melbourne.

Radford, L., Edwards, L., & Arzarello, F. (Eds.). (2009). Gestures and multimodality in the construction of mathematical meaning [Special issue]. *Educational Studies in Mathematics, 70*(2).

Rosch, E. (1999). Reclaiming concepts. In R. Nunez & W. J. Freeman (Eds.), *Reclaiming cognition: The primacy of action, intention and emotion* (pp. 61–78). Thorverton, England: Imprint Academic.

Roth, W.-M. (2001). Gestures: Their role in teaching and learning. *Review of Educational Research, 71*(3), 365–392.

Roth, W.-M. (2010). Incarnation: Radicalizing the embodiment of mathematics. *For the Learning of Mathematics, 30*(2), 8–17.

Roth, W.-M., & Thom, J. S. (2009). Bodily experience and mathematical conceptions: From classical views to a phenomenological reconceptualization. *Educational Studies in Mathematics, 70*(2), 175–189.

Rotman, B. (2009, September). Gesture, or the body without organs of speech. *Semiotix. A Global Information Bulletin.* Issue 15. Available at http://www.semioticon.com/semiotix/semiotix15/sem-15-02.html

Varela, F., Thompson E., & Rosch, E. (1991). *The embodied mind.* Cambridge, MA: MIT Press.

Vygotsky, L. S. (1997). *Obras escogidas V–Fundamentos da defectología [The fundamentals of defectology]* (Julio Guillermo Blank, Trans.). Madrid, Spain: Visor.

Vygotsky, L. S. (1978). *Mind in society: The development of higher psychological processes.* Cambridge, MA: Harvard University Press. (Original work published 1930)

SECTION III

GESTURE AND EMBODIMENT IN UNIVERSITY MATHEMATICS

CHAPTER 7

EMBODIED COGNITION ACROSS DIMENSIONS OF GESTURE

Considering Teachers' Responses to Three-Dimensional Tasks

Deborah Moore-Russo
University at Buffalo, State University of New York

Janine M. Viglietti
Buffalo State College, State University of New York

There has been a recent shift in the way researchers consider cognitive activity, moving away from Descartes' argument that man is a "thinking thing" for whom bodily constraints are a liability (Anderson, 2003) to recent suggestions that cognitive activity is embedded in the body's experiences (Barsalou, Niedenthal, Barbey, & Ruppert, 2003; Lakoff & Johnson, 1980; Smith, 2005). The idea of embodied cognition blurs the distinction between the mind and body by proposing that individuals' perceptions and interactions guide their

Emerging Perspectives on Gesture and Embodiment in Mathematics, pages 153–173

conceptual and communication structures (Lakoff & Johnson, 1999). Edwards (2011) has suggested that "[t]he mind is an on-going cumulative trace, in the form of neural patterns, of experienced and imaged action" (p. 299). In addition to the mind and body, the environment (including physical, social, and cultural aspects) plays a role in cognition.

Embodied cognition suggests that what happens in the mind is dependent on (a) the task (goal-directed action) at hand, (b) the physical movements of the body, (c) how the body is situated in and interacts with the environment,[1] and (d) the peculiarities of the environment in which the body is situated. Johnson (2007) stated,

> Meaning... arises through embodied organism-environment interactions in which significant patterns are marked within the flow of experience. Meaning emerges as we engage the pervasive qualities of situations and note distinctions that make sense of our experience and carry it forward. The meaning of something is its connections to past, present, and future experiences, actual or possible. (p. 273)

Thus, mind, body, and environment all work together in a synergistic system.

The mind and the environment impact bodily actions, including the gestures a person uses when reasoning and communicating. The objects found in the environment and how the body interacts with them influence thought. Finally, the mind and body work in unison to change the environment and how it is perceived. For example, when a person sketches a drawing, a new artifact is introduced into the environment, which in turn feeds back into the way the person thinks about what it is he or she has tried to represent in the drawing. Embodied cognition involves the co-evolution of the mind, the body, and the environment as an individual engages in tasks. Each of the three undergo constant modification, development, and refinement, with any one often adapting to accommodate changes in the others.

DIMENSIONS OF GESTURES

Gestures involve placement or movement of a part, or multiple parts, of the body. Individuals most commonly use their hands or arms when gesturing. Gestures can be spontaneous or deliberate actions. Individuals may be unaware that they are using gestures, or they may purposely use gestures, for example, omitting certain words and replacing them with particular gestures.

Individuals use gestures to communicate ideas to others as well as to themselves. In communication with others, gestures are often used to display emotion, steer attention, emphasize a point, or convey information. In this chapter, we will focus specifically on gesture or what Kendon (1980, 1993) classified as gesticulation (i.e., a motion of a part of the body that conveys a meaning other than its own motion and is semantically relatable to the accompanying speech) (Kendon, 1997). To be specific, when we use the word *gesture* in this chapter, we are referring to the bodily actions that relate to the accompanying speech or reasoning of an individual.

David McNeill (1992, 2005) proposed four dimensions[2] of gesture: iconic, metaphoric, deictic, and beat. Iconic gestures directly represent some aspect of an object, action, or event found in speech. They bear a close relationship to the content of the speaker's message, the intended referent. For example, Figure 7.1 displays a common gesture often seen when an individual tries to reason about or communicate information regarding intersecting planes. This gesture involves two hands placed so that the bottom edge of one touches the top edge of the other. From the vantage of the person gesturing, the two hands appear to form an "X" shape. Although the hands do not actually intersect each other, the gesturer intends observers to imagine each hand embedded in a plane that extends infinitely.

Metaphors are used in speech as a means to help understand one concept in terms of another. Similarly, metaphoric gestures represent an observable entity or action that, through a metaphoric relationship, represents the content of the speaker's message, although it is not literally applicable to it. While iconic gestures directly depict a referent in speech, metaphoric gestures indirectly represent an object or action found in speech through a "cross-domain mapping" (Cienki & Müller, 2008). The gesture maps to a concrete, observable object or action that in turn maps to another object or action that is more abstract or beyond perception.

Metaphoric gestures can be idiosyncratic or common to a culture. For example, rotating one's hand while discussing opening a door is an example of an iconic gesture, in which the gesture mimics the action of turning the doorknob. The same hand gesture when accompanied by the expression "turned his life around" is an example of metaphoric gesture

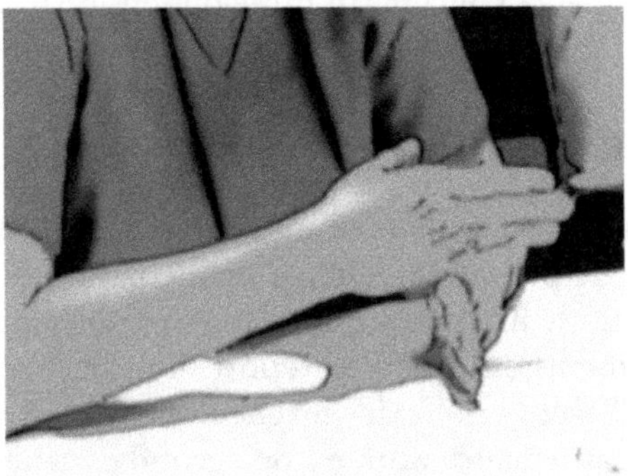

Figure 7.1 Common gesture representing intersecting planes.

accompanied by metaphoric language. The underlying metaphor maps the changing of one's life choices to spatial movement. The same gesture when accompanied by the expression "went from being a 'loser' to a productive citizen" also exemplifies metaphoric gesture, although no spatial language is employed.

Deictic gestures are pointing gestures most commonly made by extending a single finger or the entire hand, but they can also employ other body parts or even external objects (e.g., a pen in hand). Deictic gestures are often used to direct attention to an object in a real or an imagined external frame of reference. While individuals frequently point at environmental artifacts (i.e., objects that are physically present in the environment), Bühler (1982) refers to pointing at objects that are not present as *deixis at phantasma.* For example, pointing to a hat on one's head while asking a friend's opinion of it is an example of a deictic gesture. Pointing to one's head and referring to the hat worn yesterday is an example of *deixis at phantasma.*

Beat gestures are rapid, often repetitive gestures, which provide "temporal highlighting" (McNeill, 2011) to the accompanying speech of an individual. They almost always involve a speaker's hands beating out a rhythm. McNeill (2011, p. 146) describes beat gestures as "zeroing in rhythmically on the prosodic peaks of speech... [as well as] signaling the temporal locus of something the speaker feels to be important with respect to the larger context." Beat gestures are frequently seen in political speeches.

Goldin-Meadow and Beilock (2010) suggest that gestures do not simply reflect individuals' thinking; they are also capable of changing how individuals think. We offer the position that each dimension of gesture can be related to the larger body of research on embodied cognition. We apply McNeill's four dimensions to gestures as well as to bodily actions that use objects from the environment to reason and communicate. In the next section, we highlight actual responses observed when individuals engage in problem-solving tasks as a means to help unravel how individuals reason about basic mathematics concepts in three-dimensional space.

EXAMPLES

For this chapter, we draw from a corpus of data collected over a period of six years in which 81 pre- and in-service mathematics teachers were asked to work collaboratively in groups of four to five on problem-solving tasks involving three dimensions, then to report their answers individually. Of the participants, 28 were pre-service teachers with the equivalent of a bachelor's degree in mathematics working on secondary teaching certification requirements, while nine were pre-service teachers with undergraduate degrees with at least 20 credits of mathematics working on elementary teaching certification with an emphasis in mathematics. Two of the participants were undergraduates working on secondary mathematics education degrees. The remaining participants were in-service secondary mathematics teachers: 30 with less than three years teaching experience and 12 with more than three years experience teaching mathematics.

The participants were present in one of two settings: either as members of a workshop at a regional mathematics teachers' conference or as members of a graduate mathematics education course. In both settings, the teachers engaged in tasks that were intended to deepen their understanding by requiring them to explore and describe fundamental three-dimensional objects and the objects' relationships with each other. Because the activity took place at the beginning of the graduate course, teachers in both settings had about the same amount of experience working with the members of their groups. The primary difference between the two settings was that the teachers in the professional development workshop were not given an accompanying text that outlined the tasks. In this setting, the task was presented without introductory text or accompanying figures.

The initial purpose of the research was to study teachers' communication and understanding of axes in three-dimensional space. During the course of analysis, the research team observed that the teachers' development of conceptualizations of the objects and concepts of three-dimensional space was multimodal in nature. It became evident that the nature of the participants' reasoning was strongly related to their use of language, gesture, available materials, and the environment in which they worked.

While we will use multiple, related examples, we will repeatedly highlight the gestures used in the communication between two individuals from the graduate course, Alice and Betty,[3] who were working in a five-person group during an episode that lasted ten minutes and two seconds. This particular episode was selected because it provides a representative sample of the gestures we have observed.

During the episode, the five-member team was given the following text, figures, and embedded questions.[4] Particular attention will be given to the gestures used in the discussion generated by the second question below, which was followed by Figure 7.2.

> Let the *x*-axis represent east/west movement, the *y*-axis north/south movement, and add an additional axis, one that is perpendicular to the others, to define a three-dimensional system of coordinates. Let's call that third axis the *z*-axis; think of it as being above or below sea level.

Next, the text introduced the origin as the point where the three axes meet. Then it discussed ordered triples and the plotting of points in three dimensions. A second figure with the same orientation as the first was

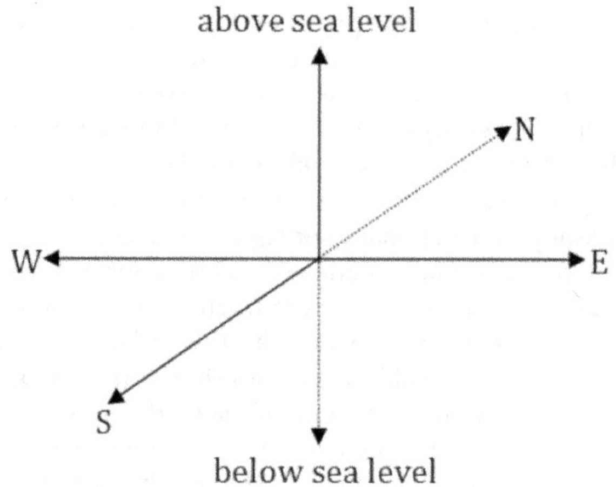

Figure 7.2 First embedded figure in text.

provided but with *x*-, *y*-, and *z*-axis labels in place of the W-E, S-N and below-above sea level labels, respectively, as seen in Figure 7.3. This figure also showed how to plot the ordered triple (–7, 4, 6).

Just after this, the first question asked the teachers to: Write the ordered triple that represents the origin in three-dimensional space. The second question that immediately followed asked: If the axes of the *xy*-plane divide it into four regions (called quadrants), how many regions do the axes of the *xyz*-space divide it into? Explain your reasoning.

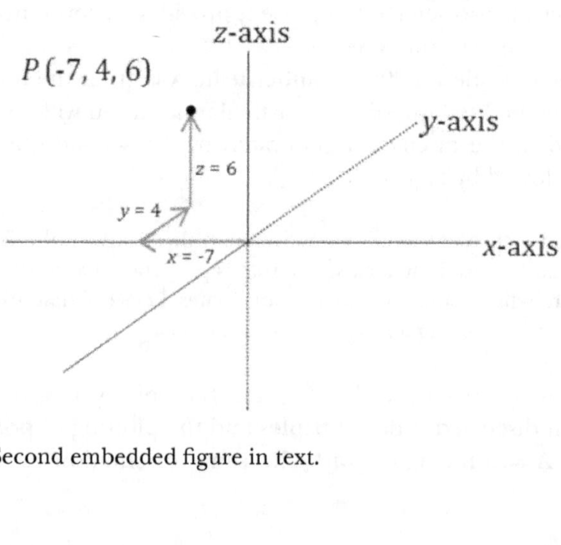

Figure 7.3 Second embedded figure in text.

ICONIC GESTURES

The teachers frequently used iconic gestures to reason about this three-dimensional task. Even when no communication with others was involved, gesturing was interpreted by the research team as a means for an individual to reason about the situation. For example, in one group, a participant read the text and stopped after the question that asked for the number of regions determined by the axes in three dimensions. She sat for a few seconds before interlacing her fingers to represent intersecting planes while pondering the situation as seen in Figure 7.4. After staring at this iconic gesture, she eventually started discussing the problem with her group members. She was only able to join the group discussion after she had created her own physical manifestation of the situation. This observation calls to mind the work of Goldin-Meadow and Beilock (2010), who found that children who used iconic gestures performed better on posttest assessments than children who did not use gesture. Although the research team does not have a measure of cognitive skills for these participants, it seemed clear that the participant needed to create a physical model before engaging with the group, and that she used her hands to create such a model.

Researchers have suggested that the synchronicity of gesture and speech is a universal phenomenon that provides a clearer understanding of cognition (McNeill, 1992). Although some iconic gestures were observed when people were engaged in noncommunicative reasoning, most of the gestures were accompanied by speech due to the collaborative nature of the setting. Even when used in communication, these gestures help untangle how the person thought about the concept.

Alice often made use of gesture and speech simultaneously. One of her first gestures is shown in Figure 7.5. She mimicked this iconic gesture after

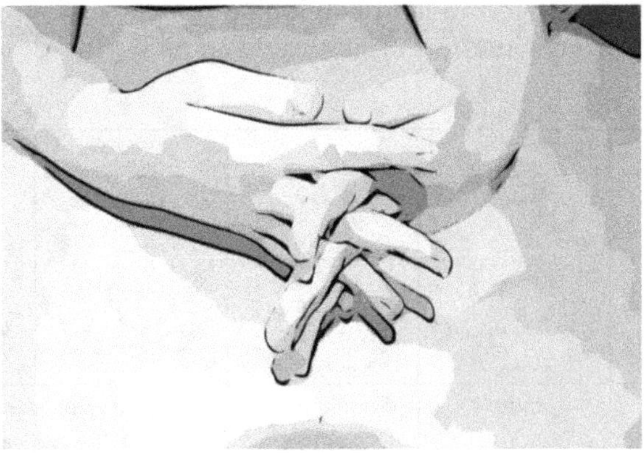

Figure 7.4 Another gesture representing intersecting planes.

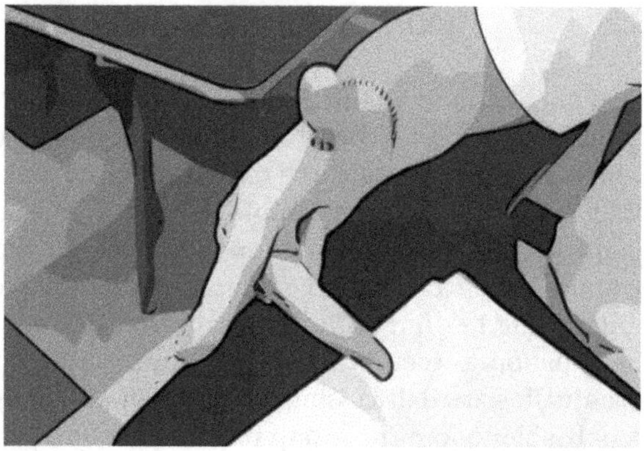

Figure 7.5 Gesture representing the *x*-, *y*-, and *z*-axes in three-dimensional space.

Chris, one of her group members, used it to discuss the orientation of the *x*-, *y*-, and *z*-axes in the text. Although the figure provided in the text was labeled, Chris chose to use an iconic gesture to talk about the axes rather than point to the paper. This suggests that it is possible that part of Chris's knowing was related to the gesture (physicality of knowing).

Later, a member of the group voiced confusion about the orientation of the axes. Having established earlier that the axes were orthogonal, Alice rotated her wrist (as shown in Figure 7.6) while explaining that as long as all three were orthogonal to each other, it did not matter where the *x*-, *y*-, or *z*-axes were located. In this iconic gesture, Alice did not try to re-create the axes but rotated her hand. Radford (2005) indicated that gestures help make evident to the gesturer salient aspects of a given concept. In this case, the rotation was salient because the three intersecting axes had already been established through prior gestures. In other words, there is a connection between gesture and developing understandings. Alice's rotation of axes through the use of gesture exemplifies the development of

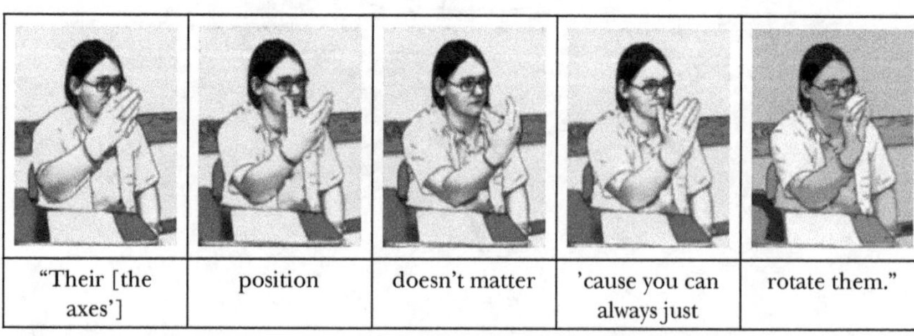

"Their [the axes']	position	doesn't matter	'cause you can always just	rotate them."

Figure 7.6 Rotating gesture used to discuss orientation of the three axes.

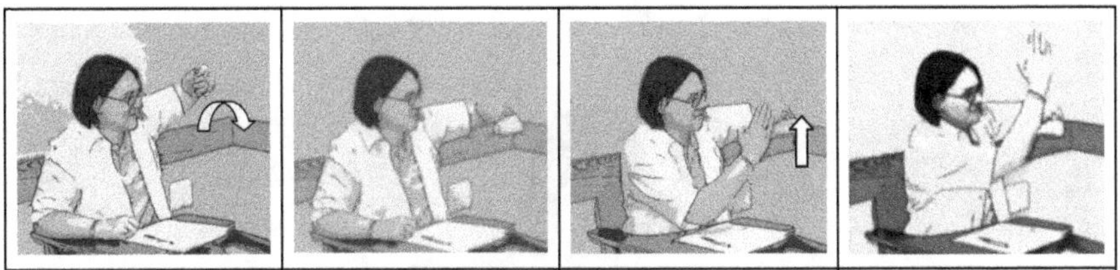

"It's like taking the *xy*-plane...[6]	and dropping it down...	then the *z*-axis	comes up."

Figure 7.7 Gesture representing the *z*-axis rising from the *xy*-plane.

understanding as she negotiated conceptualization of the orientation of the axes through the act of the rotation.

In the following example, Alice was trying to help Betty understand three-dimensional space. At one point, after talking about the three axes intersecting, Alice faced Betty and spontaneously gestured as shown in Figure 7.7.[5] Alice used a series of gestures to describe the *z*-axis rising out of the *xy*-plane. In the first image, Alice represented the *xy*-plane by holding her hand flat with her palm out in Betty's direction. The second image depicts how Alice allowed her *xy*-plane to drop and become parallel with the desktop. Finally, in the third and fourth images, Alice raised her right arm moving her representation of the *z*-axis up to, then through, the *xy*-plane. Realizing that Betty was struggling to visualize three-dimensional space (despite the presence of the figure in the text), Alice used gesture to connect a more familiar concept (the two-dimensional plane) to a more foreign concept (three-dimensional space).

While the figure in the text represented a static formal mathematical conceptualization of the concept of three-dimensional space, the gestures that Alice used to help Betty were dynamic and based on intuitive conceptualizations of three-dimensional space. This type of negotiation between personal and formal conceptualizations that Alice displayed exemplifies Abrahamson's (2004) notion of embodied spatial articulation.

Alice made use of resources other than gesture during the episode. Continuing to explain the number of regions in three-dimensional space in terms of the *xy*-plane to Betty, Alice drew the *x*- and *y*-axes on a page in her spiral notebook and then used her pen to represent the *z*-axis "coming up" from the *xy*-plane as seen in Figure 7.8. She maintained the same orientation of the three axes as she had used in the previous gesture. In this case, Alice manually manipulated easily accessible objects in her immediate environment to represent the situation.

When confronted with the second question from the same text, many collaborative groups incorporated easy-to-reach objects to help them reason about the situation. In Figure 7.9, two of the teachers spontaneously

Figure 7.8 Artifacts representing the z-axis rising from the xy-plane.

used their folders to create an iconic figure of intersecting planes in three dimensions (left) while another used sheets of paper to model the situation (right). Many made iconic figures to represent the intersection of the planes. The most common artifacts used to represent the planes were those found easily within reach. Pieces of paper or notepads often represented planes while pens and pencils were frequently used to represent the z-axis. Many groups made use of materials in the environment as a way to support their understanding of the situation. It was often the case that teachers manipulated these three-dimensional models by rotating and moving the representations of axes and planes within the model. These off-loading

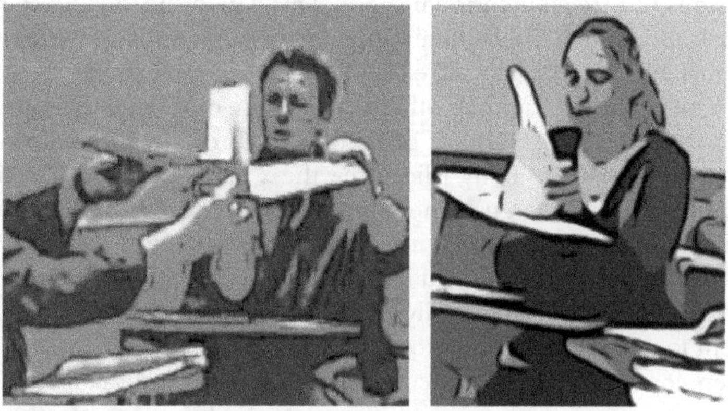

Figure 7.9 Artifacts representing intersecting planes.

strategies may have allowed the teachers to allocate fewer cognitive resources to imagining a model of the situation and more resources to working within the model. These models served more than a communicative purpose. For many groups, they were a means to facilitate cognitive activity as has been seen with respect to gesture (Iverson & Goldin-Meadow, 1998).

In both settings, at least one member of each group used gestures or gestures involving environmental artifacts iconically to represent the situation. As previously mentioned, the second question regarding the number of regions formed in three-dimensional space by the *x*-, *y*-, and *z*-axes was given without the first problem or accompanying text or figures to small groups of mathematics teachers at a conference workshop. All but 1 of the 13 teachers in the workshop used gestures alone or with objects that were readily available to model the situation as shown in Figure 7.10. These teachers were even more likely than those who had access to the accompanying text and figures to use iconic gestures and artifacts in the environment to consider the situation. For the other 68 teachers in the graduate course, every group made use of gestures or gestures with easily accessible artifacts to represent the task; however, this occurred most often when one or two teachers explained the situation to the others. In groups where a person did not understand immediately, that person and others in the group often used gesturing and artifacts to model the situation.

In Figure 7.10, one group of teachers (left) created the intersection of the three planes using their hands to hold the sheets of paper. Another group of teachers (middle) used both paper and their hands to create the intersection of the *xy*-, *yz*-, and *xz*-planes. A third teacher (right) used pens to model the intersecting axes. This served as a catalyst for another teacher to abandon the model he had been using, intersecting papers representing planes, in favor of creating his own pen model.

In both settings, with and without accompanying text, the objects in the immediate environment seemed to impact the teachers' reasoning about the situation. The teachers used easily accessible objects to off-load cognitive activity onto the environment (Wilson, 2002). In these instances, objects in the environment (a) helped the teachers visualize the situation,

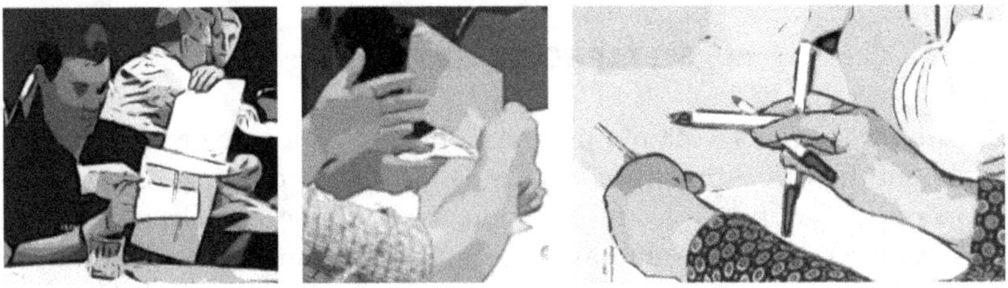

Figure 7.10 Gestures and artifacts representing intersecting planes or axes.

and (b) in many cases reason about the number of regions that would be formed. The following excerpt demonstrates one group of teachers' use of a three-dimensional model to ascertain the number of regions in three-dimensional space.

> Kara revoiced her previous statement that the three axes would create four "large" regions as she pointed to each of the four regions on the model where z values were positive. This elicited responses of disagreement from two of the members of her group. Donna picked up the model and pointed to the z-axis stating that it was positive above the xy-plane. She then used her finger to become the z-axis below the xy-plane indicating that this was where the z-axis was negative. She pointed to each quadrant of the plane as she counted out loud, indicating with a vertical sweep of her hand that those four regions were above the xy-plane and hence z was positive in this space. Interrupting her, another member of the group, Dave, took two pieces of paper and extended the x- and y- axes on the model into orthogonal planes. As he held the partitions in place he and Donna jointly explained that there would be "four regions on top and four below." Kara pointed at the xy-plane on the model and asked, "Is this a plane?" Dave confirmed, "Yeah, it's the xy-plane." Donna reoriented the plane vertically and used her hand to embody the regions where the x- and y-values were positive and negative. She then repositioned the model so that the z-axis was once again vertical.

In each of the examples above, all of the teachers were given the same goal-directed task in a collaborative setting where they were asked to work with other teachers. They all used their bodies as well as easily accessible objects particular to the environment in which they were working to model the situation. Regardless of the presence or absence of the text with its embedded figures, the groups of teachers responded similarly to the question involving the number of regions formed in three dimensions. All groups used gestures, artifacts, or a combination of both to form an iconic representation. The presence or absence of the text with its embedded figures did not impact the fact that certain individuals used their hands or sheets of paper to represent planes, while others used their fingers or pens to represent axes, to model the situation.

METAPHORIC GESTURES

Metaphoric gestures were not used as frequently as iconic gestures to reason about the three-dimensional tasks. Groups tended to use verbal metaphors only when first introduced to the three-dimensional environment. This, however, was not common; it was even less common for them to use metaphoric gestures. For example, one member in the group with Alice and Betty described positive movement on the z-axis as "climbing up a ladder."

Figure 7.11 Gesture representing three-dimensional space as a box.

Another member of this group referred to the negative z-axis as "digging down in the dirt." Neither use of metaphor was accompanied by gesture.

Alice suggested that she thought of three-dimensional space as a box. The series of pictures in Figure 7.11 illustrate the series of complex gestures Alice used to introduce the box into the environment. The initial gestures in the sequence draw out the edges of the box. In subsequent gestures, she used a pinching gesture to designate the resulting regions. The series of gestures was accentuated by sound effects that accompanied the gestures.

Chris suggested that Alice was "cutting the *JELL-O*®[7]; there are eight squares in *JELL-O*®." Alice continued with Chris's metaphor, adding gestures as she described the number of resulting regions. This series of gestures can be seen in Figure 7.12.

Gesture and speech seemed to develop simultaneously as Alice was presented with an alternate metaphor for conceptualizing the problem. She was able to instantly accept a new metaphor, the block of *JELL-O*®. Alice did not simply agree that there would be eight blocks; she also involved a physical response with her speech. This supports previous findings that gesture and speech evolve together (Bjuland, Cestari, & Borgersen, 2007). Further, Alice's combination of gesture and speech supports Reynolds and Reeve's (2001) claim that both play an integral role when conceptualizing problems. Researchers have suggested that gesture parallels linguistic use of metaphor and that the two are co-produced (Cienki, 1998; McNeill, 1992; Núñez, 2006; Sweetser, 1998).

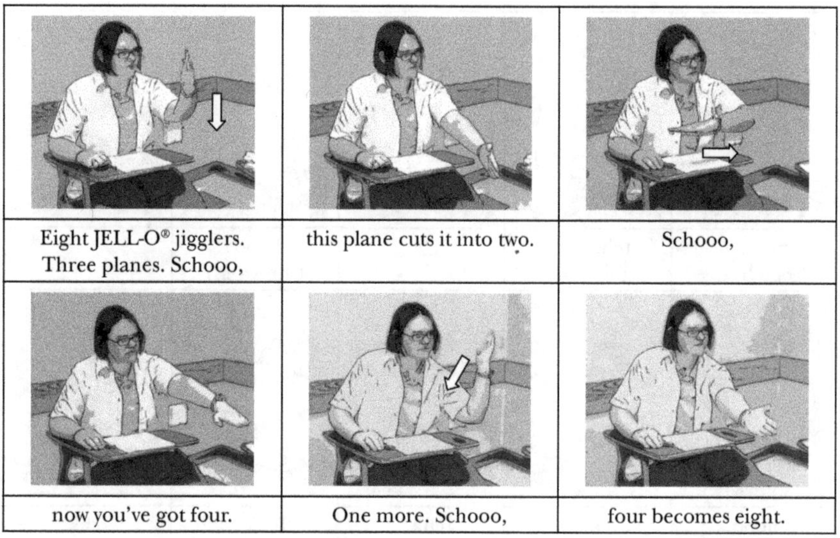

Figure 7.12 Gesture representing three-dimensional space as *JELL-O®*.

Alice's initial box metaphor emphasized the intersection of the three axes. When the *JELL-O®* metaphor was introduced into the environment, Alice then began to focus on how each plane subdivided three-dimensional space. The introduction of the second metaphor prompted Alice to consider the situation in a different light. The impact of another individual's conceptualization of the problem affected the way Alice gestured and, more important, thought about the situation.

DEICTIC GESTURES

Deictic gestures were used most often to indicate or draw attention to objects that were in the environment. On numerous occasions, this pointing was to the figures in the text, parts of the room, or inscriptions that a member of the group (often the speaker) had just created. Pointing frequently occurred as members of the group created and compared lists of all possible ordered triples (e.g., [+, +, +], [–, +, +], [+, –, +], etc.) or when an individual created a hand-drawn, two-dimensional figure representing the situation.

As part of the second question, the text referred to the four regions of two-dimensional space: If the axes of the *xy*-plane divide it into four regions (called quadrants), how many regions do the axes of the *xyz*-space divide it into? Explain your reasoning. When reading this section of text aloud, Betty spontaneously made use of deictic gesturing to point to each of the four quadrants as seen in Figure 7.13.[8] It has been suggested that gestures

Okay, sob... we have like quadrants one,	two,	three,	four.

Figure 7.13 Gesture indicating the four quadrants in two-dimensional space.

and speech are processed together (Mayberry & Jacques, 2000). Additionally, Núñez (2006) asserted that speech and gesture are two facets of the same cognitive processes. Betty's spontaneous use of gesture and speech to think about the four quadrants exemplifies such notions.

Pointing was also seen in combination with iconic use of artifacts. Each teacher in the class had been provided with a physical model, including a pegboard and pegs that could stack to consider three-dimensional space. One teacher used this model to create a list of ordered triples, which represented the eight regions. While compiling her list, the teacher extended her finger into one of the regions formed by the model as seen in Figure 7.14. While holding her finger in place, the teacher verbalized the signs of the axes to construct an ordered triple that described the region. Deictic gestures are used to draw attention to objects that are outside of the body. In this instance, the gesture off-loads work onto the environment, effectively reducing the teacher's cognitive load. By extending her body into the physical model and becoming a part of it, this teacher might have been

Figure 7.14 Gesture indicating a region in three-dimensional space.

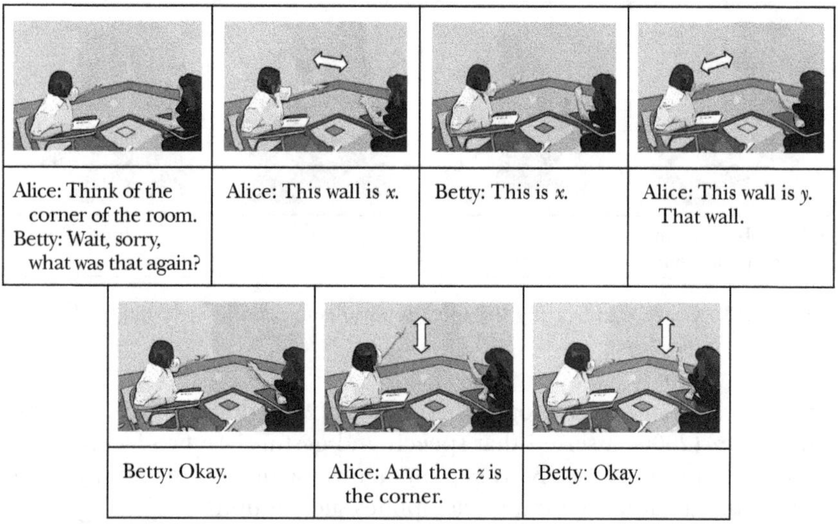

Figure 7.15 Gestures indicating the *x*-, *y*-, and *z*-axes in three-dimensional space.

saving cognitive work that would have been expended computing a solution in her head (Wilson, 2002).

A different type of pointing that occurred in numerous groups involved one member pointing to the corner of the room at the intersection of two walls using it as a metaphor to represent the three axes as seen in Figure 7.15. Alice compared each of the chair rails[9] to the *x*- and *y*-axes while she compared the seam or intersection of the two walls to the *z*-axis. The cognitive activity occurring in this episode is situated within the context of a real classroom. In this case, the gesturing was deictic while the use of artifacts in the room to model the situation was metaphoric, as noted in the language in Figure 7.15.

It has been suggested that gesture facilitates thoughts about information that is present in a given environment (Alibali & Kita, 2010). If the classroom did not have the chair rails, it is possible that Alice might not have used the particular deictic gestures she selected. It might not have occurred to her to use a metaphor at all. The notion that the environment plays an integral role in cognitive activity is a foundational component of embodied cognition (Wilson, 2002).

BEAT GESTURES

Of all the dimensions of gesture, the least frequently seen was beat. Individuals tended to use this temporal highlighting to draw attention. For example, when both of Alice's hands were occupied with a gesture, she

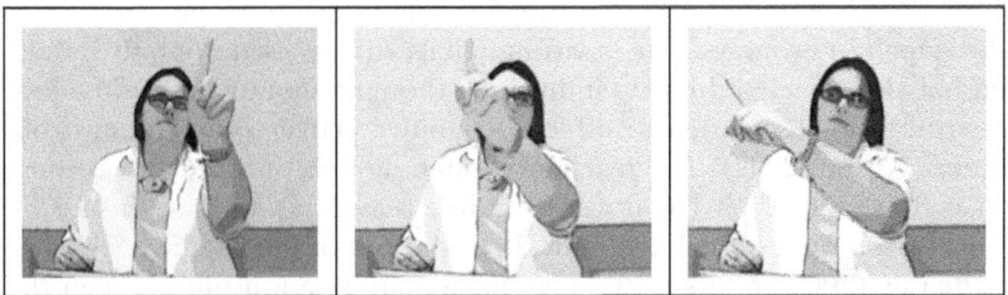

It would be two	times two	times two.

Figure 7.16 Gesture indicating an equation for the number of regions in three-dimensional space.

would hold the gesture but use a quick repetitive motion to shake the hand to which she wanted attention drawn.

In response to the second question, Alice immediately stated there would be eight regions. When one of the members of her group asked why that was the solution, Alice related the number of regions to a formula, which made use of 2^n, while using the gesture seen in Figure 7.16. In this instance, Alice's gesture highlighted information that was not perceptually present in the environment. Alice did not draw out the numerals that comprise the equation; rather, she indicated both verbally and physically through repetitious gesture (a tapping motion of her finger) the structure of the mathematical product used to solve the problem. Her gesture suggested a "memory image of formulae" as Presmeg (1986) proposed and is reminiscent of some of the gestures related to fractions and their operations that Edwards (2009) reported as iconic-symbolic in her interviews with preservice elementary teachers. For Alice, the relationship between her physical motions and the structure of the equation seemed to have strong ties.

CONCLUSIONS AND DISCUSSION

In the collaborative problem-solving environment related to three-dimensional space described here, all four dimensions of gesture were observed: iconic, metaphoric, deictic, and beat. Frequently these gestures occurred in coordination with speech as well as in conjunction with artifacts found in the environment. The examples of iconic and metaphoric gesture above often introduced new ways to physically and conceptually represent the situation. The iconic use of, and deictic reference to, artifacts typically found in classroom settings exemplified the unique ways that teachers employed items found in the environment to facilitate thought about the task at hand.

The environment played a formative role as cognitive processes were developed in two ways. The environmental artifacts used not only reflected but also affected the ways individuals thought about a given situation. First, individuals tended to off-load cognitive activity onto the environment (Wilson, 2002). The teachers created models of three-dimensional space to reduce their cognitive load while reasoning. In these instances, the environment served as a critical component of the cognitive process. Hutchins (2005) refers to the complementary relationship between the internal conceptual structure and physical model as a "material anchor" in order to stress the stabilizing role of the model on the individual's understanding. Second, cognitive activity was situated. The actual classroom environment within which the teachers worked impacted their perceptions and actions. As shown in Figure 7.10, teachers made use of sheets of paper to model planes and pens or pencils to model axes.[10] They also made use of the chair rails as shown in Figure 7.15 to model axes. These events illustrate the impact the environment can have on the way that individuals consider three-dimensional space in terms of either intersecting axes or planes.

Rather than being solely a product of the mind, cognitive activity was shaped by the gestures and artifacts used by the teachers. Alice's initial response to why there would be eight regions in three-dimensional space was accompanied by a beat gesture, as shown in Figure 7.16. Responding to cues from Betty and the environment, Alice went on to use of a variety of gestures. Although standard three-dimensional figures were represented by two-dimensional drawings within the text, Alice used a series of increasingly sophisticated gestures to represent the situation in actual three-dimensional space. It is clear through her use of gesture that her actions changed the way she thought about the situation, which in turn impacted the way she conceptualized the number of regions in three-dimensional space. As shown in Figures 7.5 and 7.6, she first considered the three intersecting axes and how the orientation of these axes did not matter. Alice then started using the *xy*-plane with the *z*-axis extending from it to consider the situation, as shown in Figures 7.7 and 7.8. In Figure 7.11, she used a box to consider the three intersecting axes, but then she switched to a consideration of the regions formed on either side of each of the three planes in Figure 7.12. Finally, in Figure 7.15, she used environmental objects (the walls with chair rails) that showed both the planes and the axes in the situation.

In this chapter, we have suggested that the environment impacts the actions of the body, and the body is part of the cognitive experience. We have provided examples of gestures and artifact use to support the notion that the mind, body, and environment work together, each impacting the others. To study teaching and learning, researchers should take into account the physical environment and the bodily actions (e.g., Moore-Russo

& Viglietti, 2012) of the individuals under investigation. If one accepts this embodied cognitive perspective, there are corresponding implications for educational research as well as education.

Making an individual's body part of the learning experience and attending to the physical learning environment is critical to how and how deeply a person understands a given concept. To improve teaching and learning, educators need to be aware of the physical learning environment and how learners (and teachers) introduce gestures or make use of artifacts found in this environment. Moreover, educators should provide learners with opportunities to engage in independent and collaborative cognitive activity that stimulates the use of artifacts and bodily actions including gestures.

NOTES

1. While the sociocultural aspects of the environment are important, this chapter will not focus on these. Note that when we use the term "environment," we mean a synonym for an individual's physical environment.
2. McNeill (2005) referred to the four as dimensions of gesture because they reflect aspects or features of a gesture rather than four different types of gesture. In fact, a single gesture might simultaneously display more than one of the four dimensions.
3. Alice and Betty are pseudonyms, as is Chris. Note that all gestures involving Alice and Betty were reenacted by the research team (after at least 50 viewings of each gesture), and then illustrated, to protect the teachers' identities.
4. As mentioned above, some teachers had access to the text (with embedded figures) whereas others were provided with the second question only. Explicit mention will be made when examples are used from teachers who did not have access to the text.
5. Arrows are included in Figure 7.7 to depict the dynamic movement of the gesture.
6. The use of ellipses in a quote indicates a verbal pause.
7. *JELL-O®* is a brand of sweetened, fruit-flavored gelatin that is used as a dessert. When a block of *JELL-O®* is cut into smaller blocks, the blocks are often called "jigglers" because they jiggle.
8. Rather than pointing to the image provided in the text (see Figures 7.2 and 7.3), Betty pointed to a referent that she imagined in front of her.
9. A chair rail is a type of molding that runs horizontally on a wall at a height to keep the tops of chairs from harming or marking the wall.
10. Similarly, the use of fingers tended to correspond with discussions about axes while the use of hands tended to correspond with discussions about planes.

REFERENCES

Abrahamson, D. (2004). Embodied spatial articulation: A gesture perspective on student negotiation between kinesthetic schemas and epistemic forms in learning mathematics. In D. E. McDougall & J. A. Ross (Eds.), *Proceedings of the 26th annual meeting of the North American Chapter of the International Group for the Psychology of Mathematics Education* (Vol. 2, pp. 791–797). Windsor, Ontario: Preney.

Alibali, M. W., & Kita, S. (2010). Gesture highlights perceptually present information for speakers. *Gesture, 10*(1), 3–28.

Anderson, M. L. (2003). Embodied cognition: A field guide. *Artificial Intelligence, 149*, 91–130.

Barsalou, L. W., Niedenthal, P. M., Barbey, A., & Ruppert, J. (2003). Social embodiment. In B. Ross (Ed.), *The psychology of learning and motivation* (Vol. 43, pp. 43–92). San Diego, CA: Academic.

Bjuland, R., Cestari, M. L., & Borgersen, H. E. (2007). Pupils' mathematical reasoning expressed through gesture and discourse: A case study from a sixth-grade lesson. In D. Pitta-Pantazi & G. Phillippou (Eds.), *European Research in Mathematics Education V: Proceedings of the Fifth Congress of European Society for Research in Mathematics Education* (pp. 1129–1139). Larnaca, Cyprus: Department of Education, University of Cyprus.

Bühler, C. (1982). The deictic field of language and deictic words. In R. J. Jarvella & W. Klein (Eds. and Trans.), *Speech, place, and action* (pp. 9–30). London, England: John Wiley & Sons.

Cienki, A. (1998). Metaphoric gestures and some other relations to verbal metaphoric expressions. In J.-P. Koenig (Ed.), *Discourse and cognition* (pp. 189–204). Stanford, CA: CSLI Publications.

Cienki, A., & Müller, C. (Eds.). (2008). *Metaphor and gesture.* Amsterdam: John Benjamins.

Edwards, L. (2011). Embodied cognitive science and mathematics. In B. Ubuz (Ed.), *Proceedings of the 35th Conference of the International Group for the Psychology of Mathematics Education* (Vol. 2, pp. 297–304). Ankara, Turkey: Middle East Technical University.

Edwards, L. D. (2009). Gestures and conceptual integration in mathematical talk. *Educational Studies in Mathematics, 70*(2), 127–141.

Goldin-Meadow, S., & Beilock, S. L. (2010). Actions' influence on thought: The case of gesture. *Perspectives on Psychological Science, 5*, 664–674.

Hutchins, E. (2005). Material anchors for conceptual blends. *Journal of Pragmatics, 37*, 1555–1577.

Iverson, J. M., & Goldin-Meadow, S. (1998). Why people gesture when they speak. *Nature, 396*, 228.

Johnson, M. (2007). *The meaning of the body: Aesthetics of human understanding.* Chicago: University of Chicago Press.

Kendon, A. (1980). Gesticulation and speech: Two aspects of the process of utterance. In M. R. Key (Ed.), *The relationship of verbal and nonverbal communication* (pp. 207–228). The Hague, the Netherlands: Mouton.

Kendon, A. (1993). Human gesture. In K. R. Gibson & T. Ingold (Eds.), *Tools, language and cognition in human evolution* (pp. 43–62). Cambridge: Cambridge University Press.

Kendon, A. (1997). Gesture. *Annual Review of Anthropology, 26,* 109–128.

Lakoff, G., & Johnson, M. (1980). *Metaphors we live by.* Chicago, IL: University of Chicago Press.

Lakoff, G., & Johnson, M. (1999). *Philosophy in the flesh: The embodied mind and its challenge to Western thought.* New York, NY: Basic Books.

Mayberry, R., & Jacques, J. (2000). Gesture production during stuttered speech: Insights into the nature of gesture-speech integration. In D. McNeill (Ed.), *Language and Gesture* (pp. 199–214). Cambridge: Cambridge University Press.

McNeill, D. (1992). *Hand and mind.* Chicago: University of Chicago Press.

McNeill, D. (2005). *Gesture and thought.* Chicago: University of Chicago Press.

McNeill, D. (2011). Gesture: A psycholinguistic approach. In K. L. Meinken (Ed.), *Encyclopedia of language and linguistics* (3rd ed., pp. 143–164). Oxford: Elsevier.

Moore-Russo, D., & Viglietti, J. M. (2012). Using the K5Connected Cognition Diagram to analyze teachers' communication and understanding of regions in three-dimensional space. *Journal of Mathematical Behavior, 31*(2), 235–251.

Núñez, R. (2006). Do *real* numbers really move? Language, thought, and gesture: The embodied cognitive foundations of mathematics. In R. Hersh (Ed.), *18 unconventional essays on the nature of mathematics* (pp. 160–181). New York, NY: Springer.

Presmeg, N. (1986). Visualization in high school mathematics. *For the Learning of Mathematics, 6*(3), 42–46.

Radford, L. (2005). Why do gestures matter? Gestures as semiotic means of objectification. In H. Chick & J. Vincent (Eds.), *Proceedings of the 29th Conference of the International Group for the Psychology of Mathematics Education* (Vol. 1, pp. 143–145). Melbourne: University of Melbourne.

Reynolds, F. J., & Reeve, R. A. (2001). Gesture in collaborative mathematics problem-solving. *Journal of Mathematical Behavior, 20*(4), 447–460.

Smith, L. B. (2005). Cognition as a dynamic system: Principles from embodiment. *Developmental Review, 25,* 278–298.

Sweetser, E. (1998). Regular metaphoricity in gesture: Bodily-based models of speech interaction. In *Actes du 16e Congrès International des Linguistes.* Paris: Elsevier.

Wilson, M. (2002). Six views of embodied cognition. *Psychonomic Bulletin & Review, 9*(4), 625–636.

CHAPTER 8

THE ROLE OF CONSCIOUS GESTURE MIMICRY IN MATHEMATICAL LEARNING

Caroline Yoon
Michael O. J. Thomas
The University of Auckland

Tommy Dreyfus
Tel Aviv University

When people converse, they often mimic each other's gestures (Kimbara, 2006). In many instances, this gesture mimicry is spontaneous and unconscious, just as people often perform and interpret gestures without consciously monitoring them (McNeill, 2005; Parrill & Kimbara, 2006). In this chapter, we examine the conscious and deliberate use of gesture mimicry in mathematics learning. Specifically, we investigate the question: How can gesture mimicry enable one to (a) understand the speaker's mathematical thinking, and (b) appropriate the mathematical meanings associated with the speaker's gestures (i.e., the grounded blends) as semiotic resources (Arzarello, Paola, Robutti, & Sabena, 2009) for developing one's own mathematical thinking.

Emerging Perspectives on Gesture and Embodiment in Mathematics, pages 175–195

We begin by reviewing research that describes the social, affective, and cognitive functions of gesture mimicry. Next, we describe the theoretical construct of catchments (McNeill, 2005) and the theory of grounded blends (Liddell, 1998, 2000) that guide our interpretations of gesture mimicry. We then present a series of vignettes where two participants (Ava and Noa) produce and mimic each other's gestures while working on a sequence of calculus activities. We show how Ava mimicked the gestures that Noa performed as a way of understanding Noa's grounded blends, and how this mimicry helped Ava appropriate the gestures and grounded blends as semiotic resources for developing her own mathematical thinking. Finally, we discuss the role of gesture mimicry as a tool for developing student understandings of mathematical ideas in instructional settings.

LITERATURE REVIEW: GESTURE MIMICRY

In conversation, people often mimic nonverbal behaviors such as posture, facial expression, and foot tapping, which largely communicate emotional and attitudinal information such as nervousness, boredom, and interest (Chartrand, Maddux, & Lakin, 2005). This chapter focuses on the mimicry of gestures that are produced together with speech and are more closely linked to the semantic content of the accompanying speech than these other nonverbal behaviors (Parrill & Kimbara, 2006). We adopt Kimbara's (2006) definition of gesture mimicry: "the recurrence of the same or similar gesture across speakers" (p. 41). In practice, the interlocutor's gestures are rarely an exact duplicate of the previous speaker's gestures. Therefore, gesture mimicry is operationally defined as occurring when the interlocutor creates a gesture that contains a subset of the gestural features (such as handshape, movement, and orientation) of the previous speaker's gestures (Kimbara, 2006). Gestural mimicry usually (but not always) occurs soon after the original gesture is performed.

Research suggests that one of the main functions of gesture mimicry is to establish rapport between speakers. De Fornel (1992) characterises the "return gesture" as a way of indicating engagement in the conversation by acknowledging the speaker's talk. Gesture mimicry can be used to convey acceptance of the speaker's view or demonstrate a developing understanding of the speaker's semantic content (Holler & Wilkin, 2011). Gesture mimicry can also be used to establish a common understanding between two people as they create a joint narrative for a third person (Kimbara, 2006).

In the above instances, gesture mimicry typically arises spontaneously and unconsciously. However, gesture mimicry can also be employed consciously and deliberately as a tool for establishing a common understanding. McNeill (2008) describes one such instance in the deliberate use of gesture mimicry as a coding tool for gesture researchers. When viewing seemingly ambiguous gestures in video data, gesture researchers often mimic the gestures as a way of embodying their thought-language-hand links. McNeill (2008) argues that a speaker re-creating another's gesture enable researchers to approximate the semantic meaning that the speaker intended; he describes this kind of mimicry as a "tool for recovering the workings of another's mind" (p. 11).

Teachers can also employ gesture mimicry in a teaching strategy called the "semiotic game" (Arzarello et al., 2009). In a semiotic game, the teacher and students mimic each other's gestures, speech, and other signs as a way of establishing and appropriating culturally shared meanings. For example, a teacher may mimic a student's gesture of a mathematical concept while rephrasing the student's speech using more precise mathematical language than the students used. The students may then mimic the teacher's gesture together with the precise mathematical language, thereby developing their institutional knowledge. Arzarello et al. (2009) describe how teachers may use this strategy consciously and deliberately in their classrooms, thereby leveraging the students' signs to establish mathematically precise knowledge.

Thus, gesture mimicry can arise spontaneously and unconsciously as a form of social interaction, and it can be used deliberately and consciously to interpret semantic information in another's gesture or introduce mathematically precise terminology. In this chapter, we present a case study of conscious gesture mimicry that occurs between two students. This case study is similar to the semiotic game in its purpose of building common understandings, but it is different in its social setting; we consider this to be a semiotic game between peers.

THEORETICAL FRAMEWORK: GESTURE CATCHMENTS AND GROUNDED BLENDS

Catchments

McNeill (2005) uses the term *catchment* to refer to a set of gestural features that recur in gestures during a segment of discourse. He defines a catchment as "a kind of thread of consistent visuo-spatial imagery running through a discourse segment that provides a gesture-based window into discourse cohesion" (McNeill, 2005, p. 116). Operationally, a catchment is identified when gestures (not necessarily consecutive ones) performed within a discourse segment are observed to have partially or fully recurring gestural features, including shape, movement, use of space, orientation, and the particular body part being used. Thus, gesture mimicry involves a catchment structure shared across speakers.

Catchments in discourse segments are used to identify conceptual themes within the speaker's discourse. McNeill (2005) argues that the recurring features in a catchment indicate the presence of linkages in the semantic content of the speaker's thoughts. Catchments can be used to identify speech and gestures related to content that is linked in the speaker's mind, thereby enabling researchers to identify meanings that the speaker considers to be similar, related, and isolated. In this chapter, we use catchments to analyze how two speakers structure their understanding of mathematical concepts related to the antiderivative.

Grounded Blends

The theory of grounded blends (Liddell, 1998) proposes that a person's gestures become conceptually significant when the person blends his or her perceived physical gesture features with other conceptual features. In the language of grounded blends, the term *Real Space* refers to "the mental representation of the physical elements of one's immediate physical environment" (Liddell, 2000, p. 342). A person's Real Space is similar to a person's physical gesture space, which is the space in front of the body in which gestures are typically performed, which usually extends vertically from the waist to the eyes and horizontally between the shoulders (McNeill, 1992, p. 86). Elements of Real Space also have corresponding physical locations, enabling them to have gestures directed toward them. However, a person's Real Space is his or her mental representation of those physical elements, not the physical elements themselves.

Liddell proposes that when people use gestures to convey meaning, they blend elements of their Real Space with elements in a second mental space to produce a new mental space that underlies their gestures. For example, Figure 8.1 shows an example of a grounded blend underlying the use of a hand gesture that represents a man speaking. The elements in the Real Space include the shape of the hand, the orientation of the hand, and the motion of the hand. These are mapped to elements in the second mental space. The shape of the hand is mapped to the shape of the man's mouth, the orientation of the hand is mapped to the direction the man is facing while talking, and the motion of the hand is mapped to the motion of the man's mouth as he talks. The resulting grounded blend is a new mental space that enables one to interpret the hand as representing the man's mouth.

In this chapter, we use the theoretical construct of a catchment to identify recurring gestural features that appear to signal a cohesive set of mathematical ideas. We then use the theory of grounded blends to map the gesture features in various catchments to precise mathematical concepts.

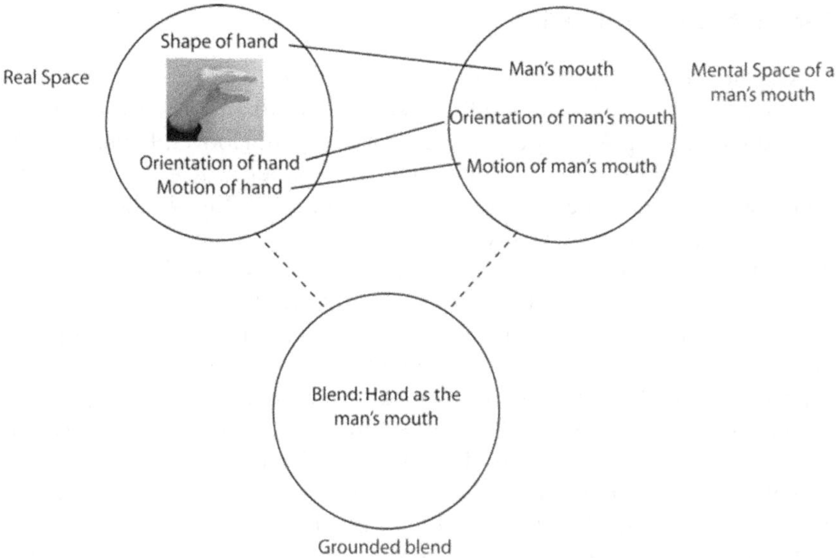

Figure 8.1 A grounded blend that gives rise to a hand being interpreted as a speaker's mouth.

METHODOLOGY

Participants

The participants in this study were two female secondary school mathematics teachers: Ava had seven years of teaching experience, and Noa had three years of teaching experience at the time of the study. Both teachers had studied calculus at the university undergraduate level, but neither had taught it at the Year 13 level (the last year of secondary school in New Zealand), as they did not feel confident about teaching calculus. In this study, Ava and Noa functioned as students working on calculus tasks for their own professional development rather than as teachers teaching in a classroom.

Ava and Noa were recruited to pilot four calculus tasks that were used in a larger study of ten undergraduate students' construction of knowledge about graphical antiderivatives. The study did not initially intend to investigate the participants' use of gestures but sought to investigate students' consolidation of calculus concepts. Consequently, the participants were not specifically instructed to use gestures while working on the activities. Nevertheless, Ava, Noa, and six of the ten undergraduate students who participated in the study spontaneously used gestures to enhance their understandings of graphical antiderivatives.

The Mathematical Tasks

Ava and Noa worked in pairs together on four calculus tasks in four one-hour sessions over the course of two weeks. The first task was created using Model Eliciting Activity design principles (Lesh, Hoover, Hole, Kelly, & Post, 2000). Accordingly, it begins with a newspaper article that describes how a person died while hiking (called "tramping" in New Zealand) a dangerous track because he was misled by the track's vague difficulty rating. Afterward, participants are given warm-up activities in which they calculate the steepness (gradient) of a given height graph of a hiking track and then sketch the gradient graph (derivative) of the track. This sets up the problem statement, which asks participants to design a method that can be used to find the distance-height graph of any hiking track from its gradient graph and to use this method to sketch a graph and give a detailed description of a specific track whose gradient graph is given in Figure 8.2. This task is mathematically equivalent to creating instructions for finding the graphical antiderivative of a function presented graphically.

The remaining three tasks also required participants to sketch and reason about graphical antiderivatives. Task 2 focused on interpreting, in terms of the features of the graph of the antiderivative, positive and negative regions of a gradient graph, as well as its maximum and minimum points. It presented functions graphically (see Figure 8.3 for examples) so as to

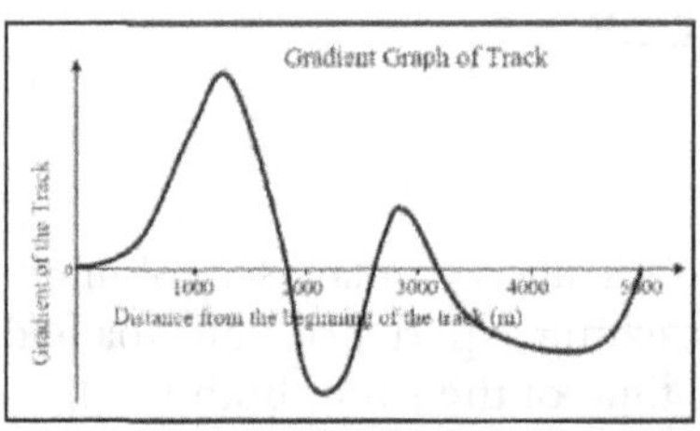

The O'Neills need your help!

Design a method that the O'Neills can use to sketch a distance height graph of the original track (like the one given in the warm up question). You can assume that the track begins at sea level.

Write a letter to the O'Neills explaining your method, and use your method to describe what the tramping track will be like on the day. In particular, you must clearly show any summits and valleys in the track, uphill and downhill portions of the track, and the parts of the track where the slopes are steepest and easiest.

Most importantly, your method needs to work not only for this tramping track, but also for any other tramping track the O'Neills might consider.

Figure 8.2 The initial activity (tramping MEA).

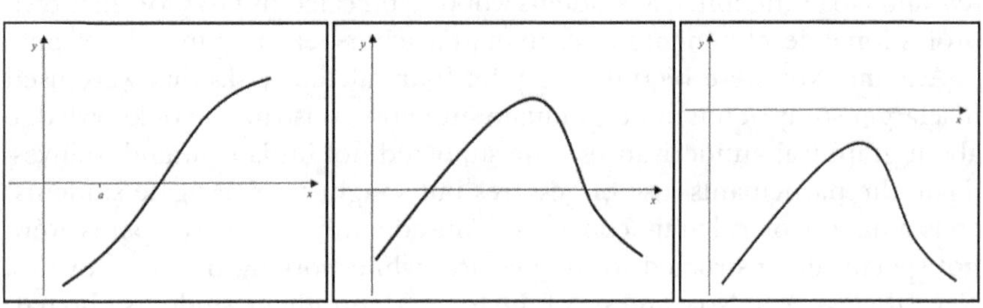

Figure 8.3 Examples of functions presented in Task 2. Students are asked to draw a graph of an antiderivative of each of these functions.

Jenni: I think that they can have a common antiderivative function.

Ray: No, that's not true, they must have different antiderivative functions because the functions are different.

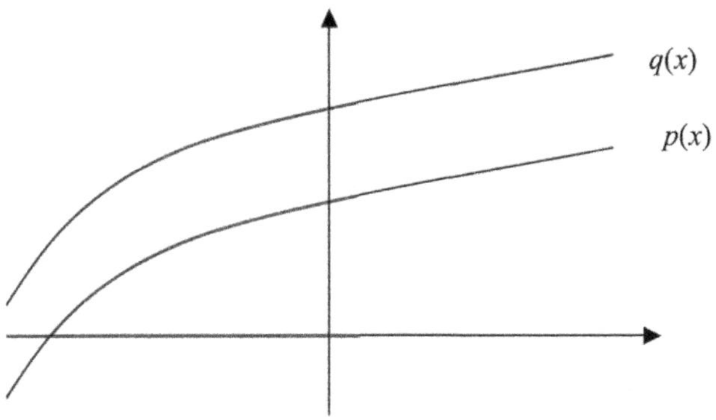

Figure 8.4 An example of a scenario presented in Task 3.

encourage participants to reason in the graphical representation; none of the functions was described with algebraic equations.

Task 3 focused on the vertical placement of the antiderivative (the constant of integration), as well as concavity and points of inflection. For example, students were asked to consider Jenni and Ray's reasoning in Figure 8.4 about whether the function $q(x)$ could have the same antiderivative as $p(x)$, where $q(x)$ was a vertical translation of $p(x)$.

Task 4 presented participants with a sequence of unmarked graphs, with each containing a function, the derivative of the function, and an antiderivative of the function, and asked the participants to identify which was which (see Figure 8.5).

Data Analysis

Videotapes and audiotapes of Ava and Noa working on the four tasks were transcribed and annotated with all the gestures (including deictic gestures) that Ava and Noa performed. We coded various visuospatial features of these gestures: orientation, motion, use of gesture space, gesture shape, use of body part(s) and other objects (like pens, rulers), along with accompanying speech. Next, we sought to identify catchments by comparing features across these gestures and organising them into groups that shared similar gesture features. We coded the gesture mimicry that occurred within each catchment to identify the timing, accuracy, and apparent purpose of the mimicry. Finally, we identified both the semiotic resources (such as

The diagram below shows a graph with three functions:

- A function f
- The derivative of f
- An antiderivative of f

Identify which is which.

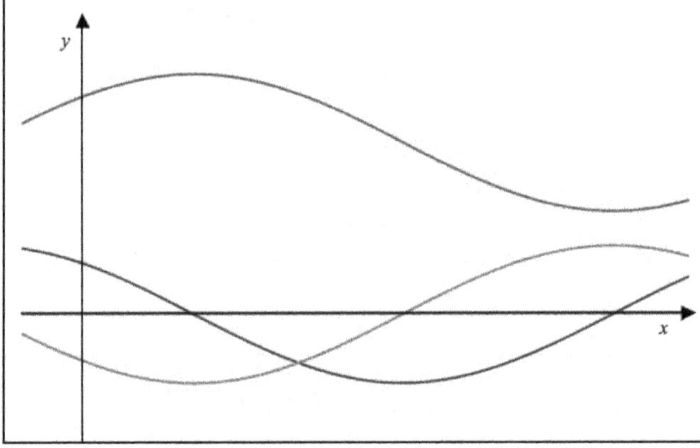

Figure 8.5 An example of a question from Task 4.

drawings, symbols, and graphs) that Ava and Noa linked with these mimicked gestures and speech, as well as the mathematical concepts that were being communicated and developed through these gestures.

THREE GESTURE CATCHMENTS AND THEIR ASSOCIATED GROUNDED BLENDS

In this section, we present and analyze three catchments in Ava and Noa's conversations. These three catchments are hierarchically related. Two of the catchments consist of subsets of the gestural features identified in a larger catchment. The gestural features in the two smaller catchments occurred in gestures that were often performed independently of those associated with the larger catchment, so we considered them to be catchments in their own right. Nevertheless, in order to acknowledge the structural hierarchy of these catchments, we call the larger catchment the primary catchment and the two smaller ones (comprising subsets of gestural features in the primary catchment) subcatchments (A and B). We present these three catchments together with our interpretation of the grounded blends that underlie the catchment gestures.

Primary Catchment

Ava and Noa used gestures from the primary catchment (an example of which is shown in Figure 8.6a) to describe their visualization of the shape and changing gradient of the antiderivative (the tramping track) of the given graph (the gradient graph of the tramping track). In these gestures, the hand is held straight (usually the left hand) and gradually changes its tilt while also moving through the gesture space in a continuous motion. We superimposed the picture in Figure 8.6a with lines and arrows to convey the dynamicity of the gestures. The thin curve shows the trace of the hand, the direction of the arrow head shows the direction of the movement, the location of the arrow head shows the location of the hand at any given moment, and the short line segment shows the tilt of the straight hand. This catchment of features first appeared in gestures performed by Noa but were mimicked and later fully adopted by Ava. These gestures were initially performed with the left hand while Noa used her right index finger to trace along the gradient graph printed on the paper.

In the grounded blend that appears to underlie gestures exhibiting this catchment of features (see Figure 8.6b), the tilt of the hand corresponds to the gradient of the tangent of the antiderivative graph at any given point. Further, the location of the hand in the gesture space is mapped to the position of the antiderivative graph, the trace of the hand's location sketches out the curve of the antiderivative graph, and the gradual changes in the tilt of the hand are mapped to the gradual changes in gradients of the

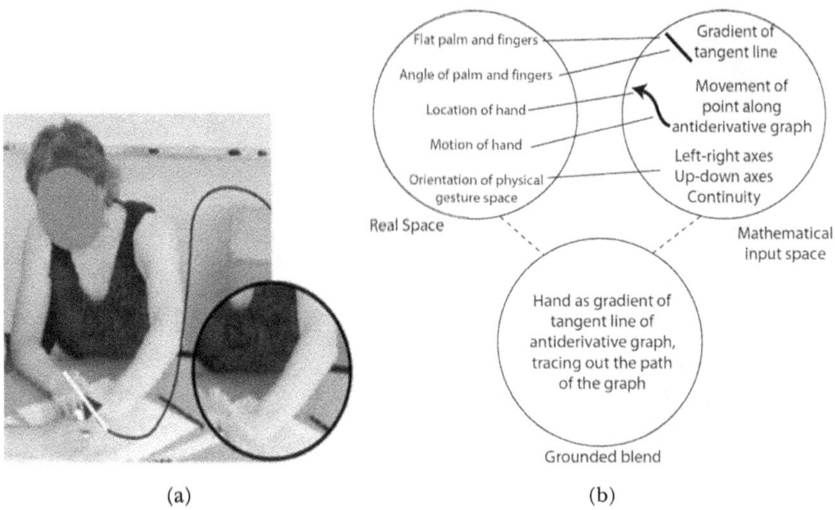

Figure 8.6 (a) Example of a gesture exhibiting gestural features in the primary catchment. (b) Grounded blend for gestures with primary catchment features.

antiderivative graph as the x-value varies continuously. Finally, the left-to-right, up-down orientation of the physical gesture space is mapped to properties associated with x–y axes in $\mathbb{R}^2$. This grounded blend is discussed in greater detail in Yoon, Thomas, and Dreyfus (2011).

Subcatchment A

Gestures that contain features found in subcatchment A were used to describe discrete gradients of points in a graph, but not the location of these points, the shape of the curve, nor the continuous changes in the gradients. In these gestures, the hand is held straight and still at a certain tilt (indicated in Figure 8.7a as a line segment) in the gesture space. In the primary catchment, the tilt of the hand changes in a gradual, continuous motion; in subcatchment A, the tilt of the hand changes in discrete intervals. The grounded blend for gestures in subcatchment A maps the tilt of the hand to the gradient of the tangent of the function at the point being considered with different tilts referring to different gradients (see Figure 8.7b).

Subcatchment B

Gestures in subcatchment B show the complement of those from subcatchment A: They are used to trace out the curve of the antiderivative graph but do not show gradients of the graph. In these gestures (shown

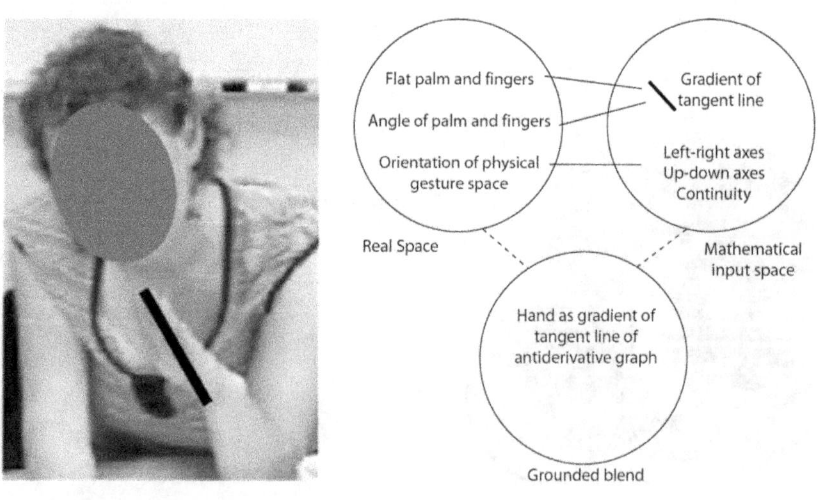

(a) (b)

Figure 8.7 (a) Example of a gesture with features in subcatchment A. (b) The grounded blend underlying gestures with subcatchment A features.

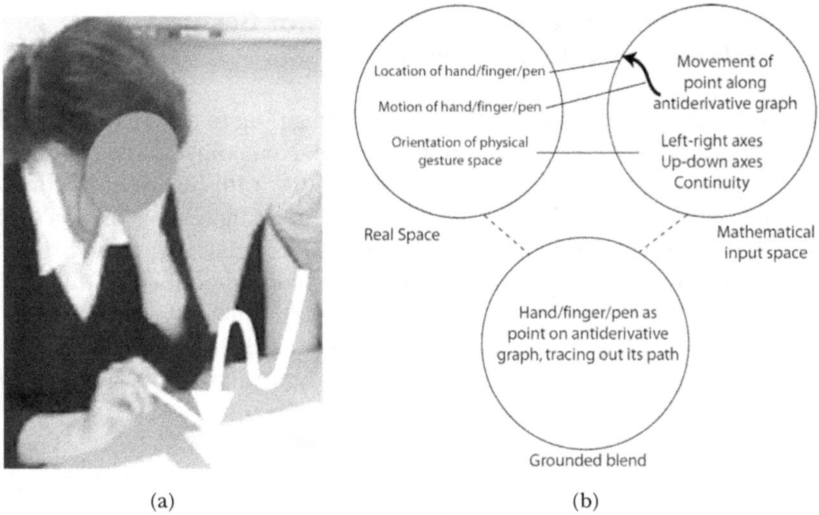

Figure 8.8 (a) Example of a gesture with features in subcatchment B. (b) The grounded blend underlying gestures with subcatchment B features.

in Figure 8.8a), the tip of a finger, hand, or pen moves within the gesture space in a sweeping, continuous motion within a plane. These gestures resemble those in the primary catchment, where the hand moves through the gesture space, tracing out the curve of the antiderivative graph, but they do not vary the angle at which the hand is tilting to represent the changing gradients of the antiderivative graph. In the grounded blend associated with subcatchment B gestures, the tip of the finger, pen, or hand is mapped to a point on the antiderivative graph, and the sweeping motion of the finger, pen, or hand is mapped to the curve of the antiderivative graph (see Figure 8.8b).

CONSCIOUS GESTURE MIMICRY

All of the three catchments described in the previous section occurred in the first three tasks, with subcatchment A also occurring in the fourth task. Because our present interest is in the mimicry of gestures as a means of appropriating them and the grounded blends as tools for thinking, we will only report on the gestures that occur in the first and second tasks. We show how, during work on these two tasks, Ava appropriates Noa's gestures and grounded blends as tools for her own thinking.

Excerpt 1: Initial Creation and Mimicry of Gestures in the Primary Gesture Catchment

Fifteen minutes into the first task, after struggling to describe via speech the shape of the track from its gradient graph, Noa points to the beginning of the gradient graph (Section A in Figure 8.9a) with her right hand and begins to gesture with her left hand (see Figure 8.9b; Noa is on Ava's left side). Noa's gestures involve all of the features associated with the primary gesture catchment and associated grounded blend described in the previous section: The angle of her hand indicates that the gradient of the track is positive and steep, the changing tilt of her hand shows that the track is becoming less steep, and the movement of her hand across space shows that the track is approaching a maximum (summit).

1-150 **Noa:** So this bit is just saying. . . . Steep hill [points to A in Figure 8.9b, gestures steep positive gradient] and still going up [points to C in Figure 8.9b, gestures gradient flattening out] but not as sharply. So really, really steep section [points to B in Figure 8.9b, gestures again a steep gradient that flattens out].

1-153 **Ava:** Steep hill [points to B in Figure 8.9b].

1-154 **Noa:** And then the point of inflection [still pointing to B in Figure 8.9b, gestures less steep gradient] and then starting to get not so steep [points to C then D in Figure 8.9b, gestures flattening gradient] up to the summit [points to E in Figure 8.9b, gestures flat gradient].

As Noa performs these gestures, Ava mimics them all the time watching Noa's gestures instead of looking at her own. This is not the first time

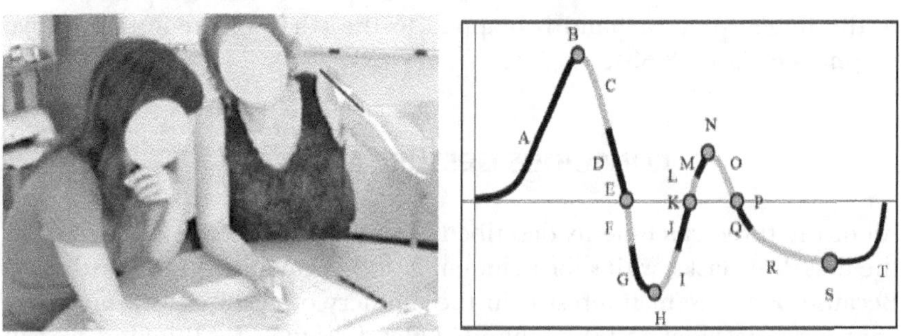

(a) (b)

Figure 8.9 (a) Noa performs gestures showing the gradient and trace of the track; Ava mimics her gestures. (b) Key sections identified in the gradient graph.

Figure 8.10 Ava mimics Noa's gestures (from the primary catchment).

Ava mimicked Noa's gestures. Earlier, Ava mimicked simple deictic gestures that Noa performed to refer to positive and negative regions of the gradient graph. When Ava begins to mimic these more complex gestures, Noa frequently looks at Ava while she continues to gesture and speak. This suggests that Noa is trying to communicate her interpretations of the track to Ava through these gestures. Ava's gestures lag two to three seconds behind Noa's, during which time she appears to check her interpretations of Noa's gestures before copying them.

1-155 **Ava:** Yes [gestures, copying Noa's gestures, to the summit].
1-156 **Noa:** And then we are starting to go downhill [points to F in Figure 8.9b, gestures negative gradient], negative gradient.
1-157 **Ava:** Yip, yip. Downhill [copies Noa's gestures of a negative gradient].

The slow speed and care with which Ava performs her gestures suggest that she is not mimicking Noa's gestures unconsciously. If she were, then she likely would have performed the gestures quicker and with more ease. Instead, Ava seems to be mimicking the iconic features of Noa's gestures consciously and deliberately as a way of embodying Noa's grounded blend. The deliberation with which Ava observes her own gestures suggests that she may be trying to perceive similar mathematical features in her gestures as Noa does in hers. In other words, Ava may be trying to construct a similar grounded blend to Noa's.

Excerpt 2: Ava Appropriates the Grounded Blend Associated With Subcatchment B

After Ava mimics Noa's gestures, she stops gesturing (at G in Figure 8.9b) and watches Noa continue to gesture her interpretation of the shape of the track. Then Ava gestures on two occasions in ways that interact with the trace of the antiderivative graph that Noa constructed through gestures (see Figures 8.11a and 8.11b).

1-158 **Noa:** And then we start going back up [points to L in Figure 8.9b, gestures positive gradient] gentle [points to M in Figure 8.9b, gestures a positive gradient] to a harder point [points to N in Figure 8.9b, gestures a less steep positive gradient. Note this is incorrect—the gesture should be steeper].

1-159 **Ava:** But not as hard as it was over there [points in the air as in Figure 8.11a].

1-160 **Noa:** But not as hard as it was over there. And then gradient is getting easier [points to O in Figure 8.9b, gestures a less steep positive gradient]. So, flattening off [Ava points as if to extend the slope Noa is indicating, as in Figure 8.11b]. We're still going up but we're flattening off.

The first occasion (Figure 8.11a) occurs after Noa remarks that the gradient of the track is becoming "harder" (lines 1-158) where she is currently gesturing, albeit incorrectly (as the gesture Noa performs is shallower than her previous gesture but should have been steeper). At this remark, Ava points to another place where she thinks Noa had previously gestured and remarks that the track is "not as hard as it was over there" (lines 1-159). Although her gesture was not physically accurate (Figure 8.11a shows that

(a)

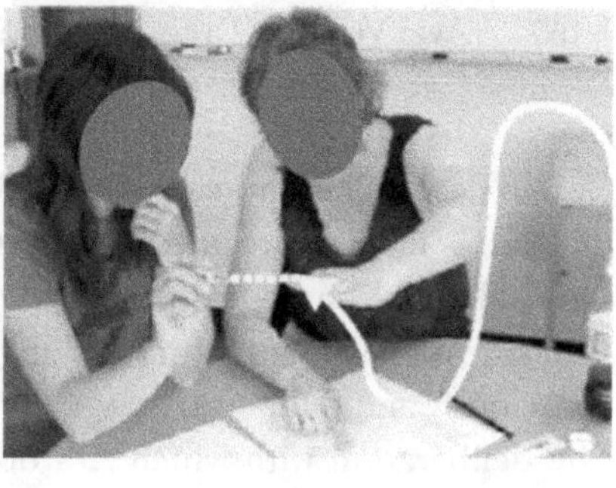

(b)

Figure 8.11 (a) Ava points to another region in the trace of the graph. (b) Ava extends Noa's gesture of the trace of the graph.

she has not located the exact spot where Noa had previously gestured a "steeper" part of the track), her gesture points toward a portion of the trace of the antiderivative graph that Noa had previously gestured. On the second occasion (Figure 8.11b), Ava extends the trace of the antiderivative graph to emphasise and agree with Noa's statement that the gradient is "flattening off" (lines 1-160).

In both of these instances, Ava performs gestures that are consistent with subcatchment B, pointing at locations within Noa's gestural trace of the antiderivative graph and extending the curve. In doing so, she goes beyond mimicking Noa's gestures and starts to contribute to the gestures. The fact that Noa and Ava appear to understand each others' gestures suggests that Ava has created a grounded blend incorporating gestures associated with subcatchment B that enables her to co-construct their interpretation of the antiderivative graph. However, it is not clear at this stage whether she has constructed the grounded blend for the primary catchment.

Excerpt 3: Ava Mimics Noa's Gestures Associated With Subcatchment A

Noa continued to use gestures with the primary catchment features throughout the first task, particularly when working out the antiderivative of the portion of the graph corresponding to Sections L-O and Q-T in Figure 8.9b. In the second task, however, she frequently performed gestures that were consistent with subcatchment A. These gestures referred to the gradient of the antiderivative but not the shape or location of the curve sketched out by the antiderivative.

2-101 **Noa:** [Traces with pen along C in Figure 8.13b] And it starts small [gestures with hand a small positive gradient], and then it gets steeper, steeper, steeper, steeper [gestures with hand a sequence of increasingly steep gradients as in Figure 8.12]. And then starts getting [Ava starts copying Noa in Figure 8.12. Noa gestures with hand a sequence of decreasingly steep gradients].

2-102 **Ava:** Oh yeah.

2-103 **Noa:** Oh no, no, no this is the change isn't it [points to B in Figure 8.13b]. So it starts at zero [gestures with hand a zero gradient] and then it starts getting steeper, steeper, steeper [traces along C in Figure 8.13b and gestures with hand a sequence of increasingly steep gradients], until it finds a constant kind of steepness.

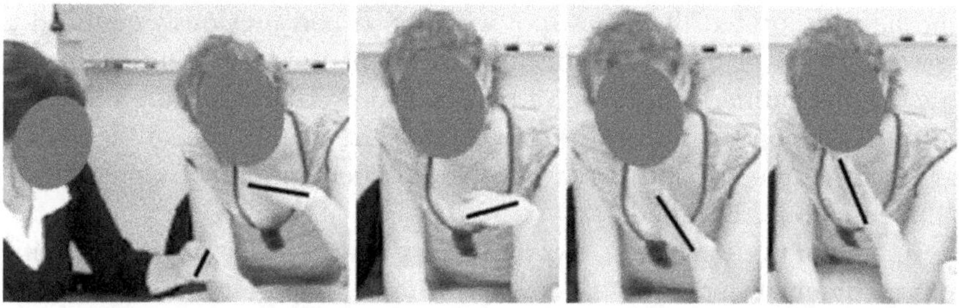

Figure 8.12 Ava mimics Noa's gestures that are consistent with subcatchment A.

Again, in this instance, Noa initiates the gestures, and Ava mimics them. Interestingly, Ava also mimics Noa's gestures that are incorrect interpretations of the gradient. Whereas Noa should have interpreted the gradients as getting steeper, she gestures and speaks of them as getting less steep. Ava appears to mimic these incorrect gestures without checking them herself, which suggests that at this stage Ava is trying to understand Noa's interpretation of the antiderivative gradients rather than using the gestures and grounded blend to guide her own interpretations. She is still engaged in mimicking Noa's gestures and hasn't appropriated the gestures and grounded blends as tools for furthering her own thinking yet. In contrast, Noa quickly discovers that she was in error and corrects herself, which indicates that she is using these tools to test and revise her own interpretations of the antiderivative gradients.

Excerpt 4: Ava Uses Primary Catchment Gestures and Grounded Blend as Tools

Excerpts 4, 5, and 6 are taken from the second calculus task, which Ava and Noa worked on three days after finishing the first task. During the second task, Ava insists on using the primary catchment gestures first to explain her interpretation of the antiderivative graph corresponding to Section A in Figure 8.13b before Noa does.

2-130 **Ava:** Ok, now let me explain it like you did. Decreasing—so here, it is very negative [points to the left of A in Figure 8.13b] which means it's very negative [makes a hand gesture indicating a very steep negative slope]. It's very steep and negative and it's getting less negative until it reaches zero [moves finger along A in Figure 8.13b and stops at B while simultaneously gesturing the slope out].

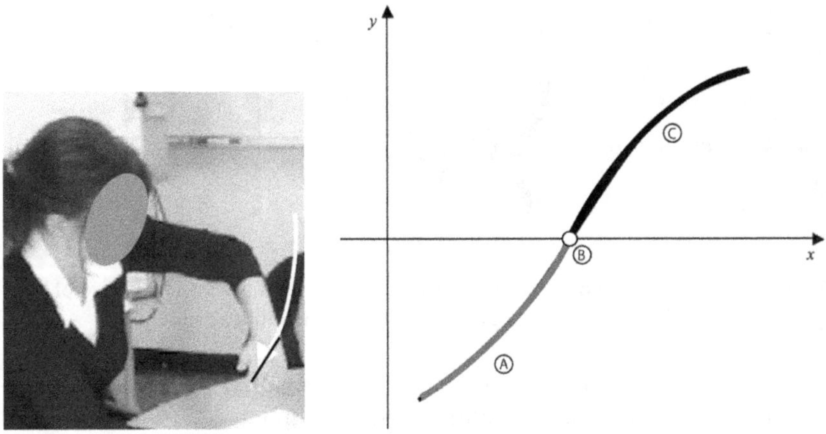

Figure 8.13 (a) Ava gestures the antiderivative graph using primary catchment gestural features. (b) The graph of the function whose antiderivative Ava is constructing.

In this instance, Ava gestures the change in the gradient of the antiderivative graph while tracing the curve. This is significant because she interprets the antiderivative graph using gestures, without Noa doing it first, which goes beyond mimicking (as in excerpts 1 and 3) or embellishing (excerpt 2) Noa's gestures. Instead, Ava appears to have appropriated the grounded blend underlying gestures in the primary catchment well enough to be able to use the gestures and grounded blend as tools for constructing her own interpretation of the antiderivative graph rather than merely trying to understand Noa's interpretation.

Excerpt 5: Coordinating Complementary Gestures

Later, when Noa and Ava try to construct an antiderivative graph of the function shown in Figure 8.3b, they use different but complementary gestures associated with the two subcatchments. The given function goes from being negative to positive and back to negative. These three portions of the given graph of the function correspond to negative, positive, and negative gradients for its antiderivative graph. Ava refers to these three portions of the given graph when she says "Negative, positive, negative." When Ava says "Negative," she gestures with the tip of her pen, tracing a continuous curve in a downward motion using the subcatchment B features (see Figure 8.14a). This becomes an upward curve when she says "positive" (Figure 8.14b) and then a downward curve when she says "negative" again

(a) (b) (c)

Figure 8.14 Ava and Noa gesture their interpretation of the antiderivative of the graph given in Figure 8.3b using gestures from the two subcatchments.

(Figure 8.14c). Noa simultaneously gestures negative, positive, and negative gradients using gestural features in subcatchment A.

In these gestures, Ava and Noa use different subsets of the gestural features in the primary catchment. Nevertheless, they appear to understand each other and see the gestures as synchronised and complementary. This suggests that Ava and Noa have both constructed grounded blends associated with subcatchments A and B, and these grounded blends may be incorporated into the overall grounded blend for the primary catchment.

Excerpt 6: Repetitive Use of Catchment as Tool

At the end of task 2, Ava and Noa attempt to construct the antiderivative of the function presented in Figure 8.3c. Whereas previous functions contained positive and negative values, this function is negative everywhere it is defined. Thus, finding the antiderivative for this function was more difficult than previous instances because it required one to consider the change in the size of the gradient rather than just considering whether it was positive or negative. In this instance, Ava uses gestures associated with the primary catchment gesturing both the trace of the graph together with the continuously changing gradients of tangents to the graph. In contrast, Noa uses gestures associated with subcatchment A, merely gesturing three discrete, static gradients associated with key changes in the gradients of the graph (very negative, less negative, very negative gradients).

2-326 **Ava:** So it's going more negative, to less negative, to more negative [gestures the shape of the curve, while Noa gestures the three gradients separately]. More negative, less negative, more negative [both Ava and Noa repeat their gestures].

2-237 **Noa:** Totally, more, less, more [Ava and Noa repeat their gestures with no speech].

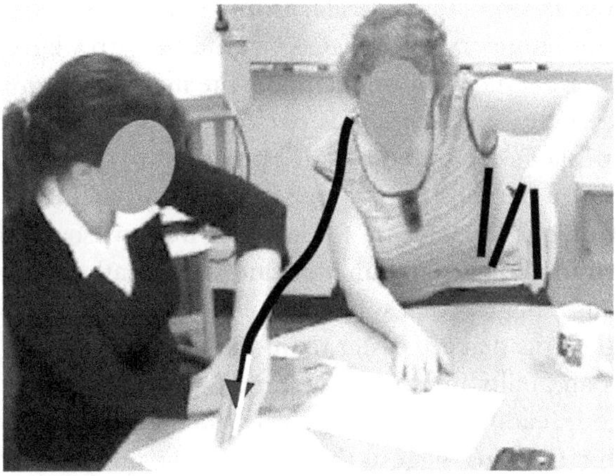

Figure 8.15 Ava and Noa repeat their gestures of the antiderivative graph three times each.

In this excerpt, Ava and Noa repeat their gestures three times each, even after they have agreed on their interpretation of the antiderivative. This suggests that they are both using the gestures and grounded blends as tools to construct, test out, and check their interpretations, not just to communicate previously constructed interpretations to each other.

DISCUSSION

Ava and Noa's case study describes a developmental progression for the process whereby an interlocutor appropriates a speaker's gestures and grounded blends as semiotic resources for enhancing his or her (the interlocutor's) own mathematical learning. In this instance, Ava began by mimicking physical features of Noa's gestures while appearing to concentrate on identifying the mathematical features that Noa had associated with them. In other words, Ava appeared to use gesture mimicry as a tool for figuring out Noa's grounded blend. Later, Ava performed gestures associated with subcatchment B: This demonstrated that she had constructed a partial grounded blend that mapped a significant subset of Noa's gestural and mathematical features associated with the primary catchment. Then in excerpt 4, Ava demonstrated that she had constructed the grounded blend associated with the primary catchment by using the gestures (and full grounded blend) to create her own interpretations of an antiderivative graph without mimicking Noa.

This case study suggests that gesture mimicry can be an effective initial step in the process of appropriating gestures and grounded blends as semiotic resources for mathematical learning. Like McNeill (2008), we propose that gesture mimicry can serve as "a kind of borrowed embodiment" (p. 10), whereby an interlocutor deciphers the mappings from the physical features of the gestures he or she is performing to the mathematical features in the speaker's grounded blend. Thus, gesture mimicry is not merely an effective way of demonstrating one's involvement and understanding in the conversation but also a means for learning how to use the speaker's gestures and grounded blends to enhance one's own mathematical learning.

The study also suggests that some complex gestures can be broken down into subcomponents that are easier to mimic separately at first. Ava initially demonstrates an ability to use subcatchment B (tracing out the shape of the curve) more readily than either subcatchment A (the gradients) or the primary catchment. This suggests that, for Ava, the mathematical challenge of tracing the shape of a curve is less complex than showing how gradients change over time. Eventually, she was able to integrate both subcatchments together in subsequent gestures (as in excerpts 5 and 6) together with the primary catchment. Thus, in this case study, gesture appropriation appears to be an iterative process, whereby subcatchments of complex gestures are initially mimicked and blended with mathematical features before being integrated with further gestural and mathematical features.

Ava and Noa's case study provides another example of the semiotic game (Arzarello et al., 2009). In this instance, however, the semiotic game takes place between peers rather than between a teacher and students. A possible practical implication could be for teachers to encourage gesture mimicry among students as they work on mathematical tasks, particularly when the students' gestures appear to be mathematically rich. When the gestures are complex, students may be encouraged to begin by mimicking subsets of the gestural features to construct partial grounded blends before connecting them with the rest of the mathematical and gestural mappings in the complete grounded blend.

REFERENCES

Arzarello, F., Paola, D., Robutti, O., & Sabena, C. (2009). Gestures as semiotic resources in the mathematics classroom. *Educational Studies in Mathematics, 70*(2), 97–109.

Chartrand, T. L., Maddux, W. M., & Lakin, J. L. (2005). Beyond the perception-behavior link: The ubiquitous utility and motivational moderators of nonconscious mimicry. In R. R. Hassin, J. S. Uleman, & J. A. Bargh (Eds.), *The new unconscious: Oxford series in social cognition and social neuroscience* (pp. 334–361). New York, NY: Oxford University Press.

De Fornel, M. (1992). The return gesture. In P. Auer & A. di Luzio (Eds.), *The contextualization of language* (pp. 159–193). Amsterdam, the Netherlands: John Benjamins.

Holler, J., & Wilkin, K. (2011). Co-speech gesture mimicry in the process of collaborative referring during face-to-face dialogue. *Journal of Nonverbal Behavior.*

Kimbara, I. (2006). On gestural mimicry. *Gesture, 6*(1), 39–61.

Lesh, R., Hoover, M., Hole, B., Kelly, A., & Post, T. (2000). Principles for developing thought-revealing activities for students and teachers. In A. E. Kelly & R. A. Lesh (Eds.), *Handbook of research design in mathematics and science education* (pp. 591–646). Mahwah, NJ: Erlbaum.

Liddell, S. K. (1998). Grounded blends, gestures and conceptual shifts. *Cognitive Linguistics, 9*(3), 283–314.

Liddell, S. K. (2000). Blended spaces and deixis in sign language discourse. In D. McNeill (Ed.), *Language and gesture* (pp. 331–357). Cambridge, UK: Cambridge University Press.

McNeill, D. (1992). *Hand and mind: What gestures reveal about thought.* Chicago: University of Chicago Press.

McNeill, D. (2005). *Gesture and thought.* Chicago: University of Chicago Press.

McNeill, D. (2008). *Gestures of power and the power of gestures.* Retrieved on March 29, 2011, from http://mcneilllab.uchicago.edu/pdfs/gsts_of_pwr_and_pwr_of_gests.pdf.

Parrill, F., & Kimbara, I. (2006). Seeing and hearing double: The influence of mimicry in speech and gesture on observers. *Journal of Nonverbal Behavior, 30,* 157–166.

Yoon, C., Thomas, M. O. J., & Dreyfus, T. (2011). Grounded blends and mathematical gesture spaces: Developing mathematical understandings via gestures. *Educational Studies in Mathematics, 78*(3), 371–393.

CHAPTER 9

RECONCEIVING MODELING

An Embodied Cognition View of Modeling

George Sweeney
Santa Ana College

Chris Rasmussen
San Diego State University

In this chapter, we take an embodied cognition perspective to reconceive the view that modeling consists of acts of translation between the domain of the real world and the domain of the mathematical world. In this reconceived view of modeling, it is unnecessary to posit the real world and the mathematical world as separate domains. Instead, mathematics (a product of mind) and the real-world phenomena coexist in the same domain. From this point of view, the body takes on a greater role, bringing together everyday and mathematical experiences. Gesture, imagination, gaze, and language are all activities used by individuals to organize their world and hence to model real-world phenomena.

Emerging Perspectives on Gesture and Embodiment in Mathematics, pages 197–226

From a classical viewpoint on modeling, Blum and Niss (1991) characterize the modeling process as an interplay between the domain of the real world and the domain of the mathematical world. The starting point for this interplay is the real-world situation, which has to be simplified or otherwise structured and then mathematized. Mathematization is the process by which "data, concepts, relations, conditions, and assumptions are to be translated into mathematics" (Blum & Niss, 1991, p. 39). The problem-solving process then "continues by work within mathematics," resulting in specific mathematical findings. These mathematical findings, which include symbolic expressions, graphs, tables, and so on, then have to be translated or interpreted in terms of the original real-world situation.

The underlying imagery for this classical view of modeling is that of cutting bonds with reality or that of detaching mathematics from its concrete embodiments (Herscovics, 1996). Figure 9.1 depicts this classical view of modeling in which mathematization and interpretation are viewed as acts of translation, either from the real world to the world of mathematics or from the mathematical world back to the real world.

Note that in this classical view, the processes of mathematization and interpretation are viewed as one-directional mappings. The domains of the mathematical world and the real world are therefore held to be separate entities in need of translation.

Rather than viewing modeling as an act of translation, we draw on Gravemeijer's (1997) suggestion to see modeling as an organizing activity. Modeling, reconceived in this way, does away with the need to treat a mathematical inscription as a result of one-directional mapping. Instead, the inscription is a record of the organizing activities in which individuals participate.

Expanding on this point of view from an embodied cognition perspective, we set forward three characteristic features that characterize modeling as an embodied organizing activity: (a) Bodily engagement organizes a particular experience by aligning seemingly different perceptual fields in space-time, facilitating the perception of an experience as mathematical in nature; (b) the use of mathematical tools shapes our expectations, expanding and constraining how we perceive a particular experience; and (c) by

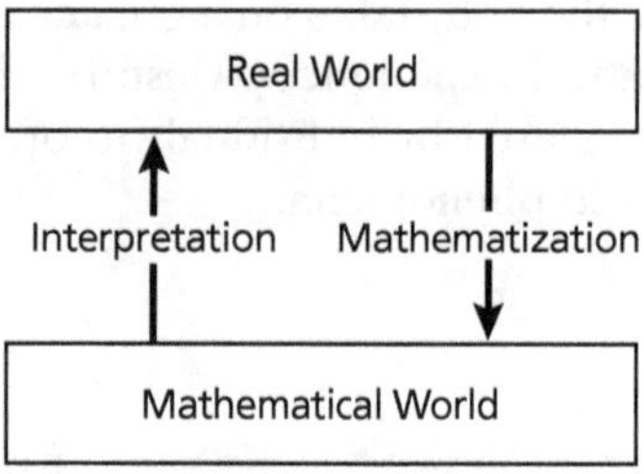

Figure 9.1 Classical view of modeling.

organizing a particular experience as a mathematical experience, new perceptions are made possible, enhancing our understanding of the original experience and facilitating new mathematics. These three characteristics set forth an alternative view on what constitutes modeling. Viewed as an embodied organizing activity, modeling therefore brings together an individual's perceptual and bodily experiences with a particular phenomenon perceived through the lens of mathematics.

THEORETICAL BACKGROUND

An important problem in analyzing the role of the body in relation to mathematical activity concerns the role of gesture and other forms of embodiment in the construction and communication of meaning. Many authors of work on gesture have assigned gesture a role of communication that helps to facilitate speech (e.g., Alibali, Kita, & Young, 2000; Alibali & Nathan, 2007; Goldin-Meadow, Nussbaum, Kelly, & Wagner, 2001). However, gesture studies such as Alibali and Goldin-Meadow (1993) have indicated that mathematically correct gestures at times precede correct speech, implying that gesture not only facilitates speech but can also precede and/or complement it. Goodwin's (1994, 2000, 2003) work has demonstrated that gesture is not isolated from context, speech, or tool use. In fact, gesture and bodily movement are always deployed as part of a multiplicity of semiotic resources working in concert as human beings construct meaning and interact in their environment (Arzarello, Paola, Robutti, & Sabena, 2009; Goodwin, 2003).

Goodwin (2003), for example, posits that bodily activity is inextricable from speech, tool use, and gesture, and that these different semiotic resources are mobilized concurrently. The concurrency of these resources therefore challenges the philosophical position that there is a thinking mind that precedes bodily activity and speech. The argument could be expressed as follows: If the mind and body are separate, then aspects of the environment are perceived by the individual and then acted on via some rational thought process, which then deploys gestures, bodily movement, and speech to express the products of these thought processes. Contemporary perspectives challenge this notion by demonstrating how gesture parses the world into culturally and interactively constructed domains that then transform the environment into relevant objects of scrutiny.

In order to capture and reflect the perspective of modeling as organizing, we reframe, at the microlevel, mathematizing and interpreting in terms of four forms of modal engagements. Hall and Nemirovsky (2011) define a modal engagement "as an *activity* someone participates in, with others, tools and symbols" (p. 3, italics added). These authors characterize the construct of modal engagement by detailing four dimensions of the body: the acting body, the encultured body, the expressive body, and the imaginary body. These four constructs constitute the primary means by which we reinterpret modeling as an embodied organizing activity.

In analysis, we highlight the acting body as a way of focusing our attention as on the individual's experience of space and the ways that tools and interaction can expand that space. An individual's sense of space is not uniformly experienced (Hall & Nemirovsky, 2011). Personal space is generally related to proprioception. Proprioception is how individuals experience their own body movements. For example, suppose that you are asked to close your eyes and extend your arm straight out in front of you. How is it that you know that your arm is in fact straight? Your proprioception lets you know that your body is in fact acting in the way that you intended (Coslett, 1998; Kennedy, Gläscher, Tyszka, & Adolphs, 2009; Vaishnavi, Calhoun, & Chatterjee, 1999). Peripersonal space is characterized by the space that is within reach of the individual. The portions of the outside environment that are directly within arm or leg length reach are sensed in peripersonal space (Holmes & Spence, 2004; Ladavas, Di Pellegrino, Farne, & Zeloni, 1998; Rizzolatti, Fadiga, Fogassi, & Gallese, 1997). Tool use that extends the reach of the individual also extends peripersonal space by allowing you to interact with objects that are outside of arm or leg length. Space that is well beyond reach is referred to as extrapersonal space, which can be experienced differently depending on the distance that an object is from an individual (Holmes & Spence, 2004; Previc, 1998).

In activity, a person pays close attention, in different ways, to each of these kinds of spaces. For example, in the act of hitting a baseball, a baseball player combines all three of these kinds of spatial sensations. He must move his feet, legs, torso, and arms towards the ball in order to strike, bringing to bear his proprioceptive senses. His body is then extended via the use of the bat and at the moment of impact the ball bringing the bat and everything within its reach into peripersonal space. He is also paying close attention to the ball in extrapersonal space as it approaches his body, as well as the peripersonal space encompassed by the reach of the bat. The coordination of all these different spatial senses allows for the hitter to hit the ball.

The second dimension, the encultured body, frames bodily activity by how the individual incorporates and utilizes tools and symbols in the process of achieving a particular set of goals. In this formulation, tools (both semiotic and physical) have a dual role. One, they extend the body into new venues for activity. Two, they bring objects into bodily reach, allowing the body to manipulate and transform objects that would otherwise not be available. The embodied nature and development of cultural tools has begun to be extensively studied in mathematics education (e.g., Arzarello, Paola, Robutti, & Sabena, 2008; Radford, 2009). Building on Hall and Nemirovsky (2011), we add to this discussion the question of how cultural resources and tools transform the experience of bodily activity. By highlighting certain aspects of phenomena and ascribing to them mathematical symbols, or by relating to a tool like a double Ferris wheel applet (online computer simulation), the individual is bringing to bear not just his or her bodily activities but also his or her full experience of these objects and their norms of use.

The third framing is called the expressive body. This framing of modal engagement includes emotional and expressive stances that the individual takes toward others or toward activity. The expressive body includes the emotional and communicative gestures that are evidenced by facial cues, body positioning, gesturing, or talk. Frequently, the impact of the expressive body is subconscious, and we, as analysts, might attribute the body's responses and actions as responses to discursive cues. For example, when a member of the class makes a comment or remark that is considered off topic or not germane to a conversation at hand, members in the immediate surroundings of the talker might shift their body positions away from the talker, indicating that they are no longer listening to what he is saying or that his contribution was not appreciated. The student could then revise his contribution or the members of his group might choose not to use what he contributed.

The final analytic framing is called the imaginary body. The perspective of the imaginary body focuses on the experience of time as it is expressed in bodily engagement. Understanding what an object or a concept is requires being able to predict the unfolding of events through time as an individual engages in that action. This knowledge of anticipated action reflects past experience with the object and represents individuals placing themselves within a set of possible future activities. Nemirovsky and Ferrara (2009) elaborate on the relationship between time and the imaginary by distinguishing between "what is" and "what could be":

> Broadening the scope of embodied cognition toward encompassing the imaginary helps us to shift our attention from "what is" to "what could be." "What is" seems the natural orientation of perceptuo-motor activity. Imagining brings up empirical or pure possibilities (i.e., "what could be") shaping the course of perceptuo-motor activity, and therefore it is particularly essential to account for the experience of abstract realms, such as mathematics. (p. 173)

Time envisioned through this lens establishes a three-part experience. On the one hand is the moment-to-moment experience of the present. The present is unfolding throughout time and is somewhat ephemeral. Experience, however, is filtered through a set of possible futures into which we are constantly moving. Our intentions and expectations for that future are given to us by our past experiences. Hence, past, present, and future are constantly interwoven into experience. Hence, as we encounter mathematical objects, we are not just experiencing them in the present moment but always considering them against a set of possible interpretations or possibilities for action. For example, in Nemirovsky, Rasmussen, Sweeney, and Wawro (2011), three girls reinvented multiplication by i as rotation. The girls did not move through multiple examples of multiplication by i, nor did they have to; rather, through their work on the floor with string, paper, and tape, they came to perceive multiplication by i across a range of possibilities where any multiplication by i would be rotation by 90 degrees. Their assessment of the multiplication by i as a rotation by 90 degrees is one that is considered in the present as a result of a particular activity, in the past as what rotation looked like and the product of multiplying by i in past mathematical experiences, and as a consideration of future action when assessing whether a multiplication by i will always be a rotation. Abstract determinations of this kind, where a person must decide that an activity will always end in a particular result, are necessarily considerations across a past, present, and future.

These four characterizations of modal engagements, the acting body, the encultured body, the expressive body, and the imaginary body, provide an analytic lens for examining mathematical cognition on a microlevel and for reframing modeling as an embodied organizing activity rather than a translating activity. These constructs enable a focus on the entirety of human experience, giving researchers a more comprehensive view of what it means for an individual to engage in mathematical modeling.

DATA SOURCES AND METHODS

The data for this report come from open-ended interviews conducted with two different pairs of students. In this chapter, we only present an analysis of one of the two pairs of students. We refer to these two students as Andrew and Oscar (not their real names). Andrew and Oscar were mathematically sophisticated, having completed the majority of their coursework in engineering and mathematics at the collegiate level. The interview tasks were designed to provide students with an opportunity to gain greater fluency with trigonometric functions. The sequence of tasks we created made use of an applet that animated a double Ferris wheel (Key Curriculum Press, 2011). The double Ferris wheel (see Figure 9.2) was represented on the computer screen by two red circular wheels (referred to as the edge wheels) connected at their centers by a blue bar (referred to as the middle bar). The edge wheels and the middle bar are all in the same plane and rotate counterclockwise. The middle bar rotates independently of the edge wheels, but the two edge wheels move at the same constant speed. The applet allows students to start and stop the rotation of the edge wheels and the middle bar or move both at the same time. The applet also gives students

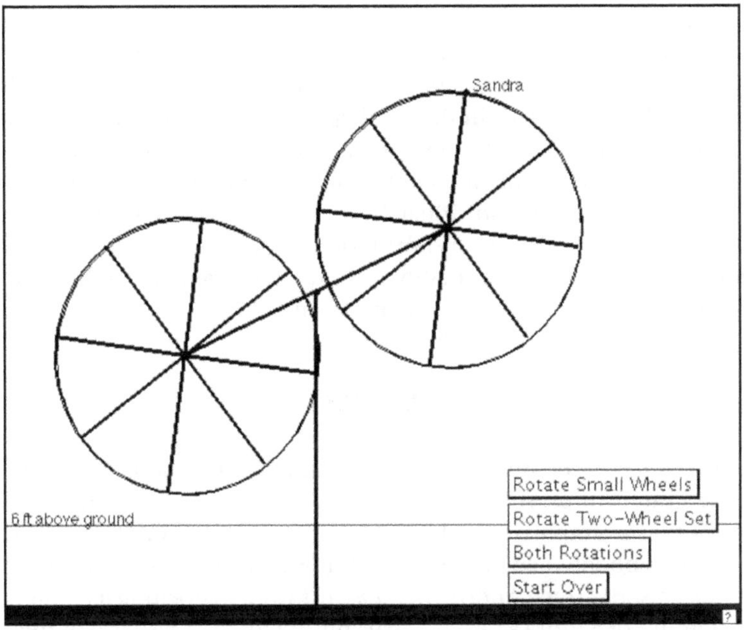

Figure 9.2 Double Ferris wheel screen shot.

the opportunity to return the double Ferris wheel to its initial starting position. Annotated on the Ferris wheel is the position of Sandra, whom the students are told is a rider on the double Ferris wheel.

The tasks that loosely structured the students' work were as follows:

1. Create a graphical representation of your choice that illustrates Sandra's ride on the double Ferris wheel
2. Create a qualitatively correct graph of Sandra's height versus time as she rides on the double Ferris wheel
3. Create a graph of Sandra's height versus time that is numerically correct
4. Create a function for Sandra's height versus time as she travels on the double Ferris wheel.

Students were given paper, markers, transparencies, tape, and other supplies to complete the tasks. For the first two tasks, the students were given only the applet and materials. After completing the first two tasks, the students were also given the radii of each of the wheels, the amount of time that it took for each of the wheels to complete a single rotation, and the distance from the ground where Sandra started. Andrew and Oscar worked together successfully to complete all of the tasks over the course of two 2-hour interviews.

Our analysis of these data employed a microethnographic approach (Nemirovsky, Rasmussen, Sweeney, & Wawro, 2011; Hall & Nemirovsky, 2011; Nemirovsky & Monk, 2000). A microethnographic analysis calls for examining short moments in the participants' activity, which allows the researcher to attend to aspects of speech, gesture, body positioning, gaze, as well as the cultural knowledge (which includes mathematical knowledge) and experience that the individual brings to bear when engaging in mathematical activity. This focus on short moments of time is necessary because cognition encompasses the totality of bodily engagement as it is situated within environments and contexts for action.

RESULTS

Our microanalysis focuses on Oscar and Andrew's creation of a height versus time graph. In presenting our analysis of this event, we focus on four episodes that exemplify how the four forms of modal engagement make possible an embodied cognition view of modeling as an organizing activity. The four episodes trace Andrew and Oscar's work in creating a qualitatively correct graph of Sandra's height versus time. This task, task two of the four-task sequence, occurred during the first session of the interviews and began just over one hour into the two-hour session.

On the first task, which invited Oscar and Andrew to create a graphical representation of their choice to illustrate Sandra's ride on the double Ferris wheel, the students chose to create a graph that related the vertical and horizontal displacement of Sandra on the double Ferris wheel (see Figure 9.3).

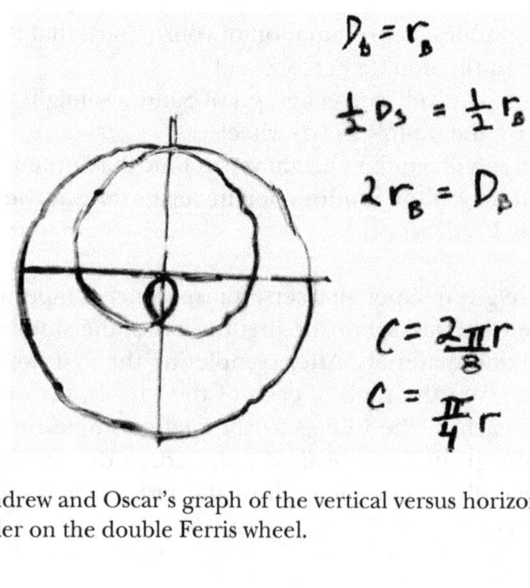

Figure 9.3 Andrew and Oscar's graph of the vertical versus horizontal displacement of the rider on the double Ferris wheel.

Figure 9.4 Andrew and Oscar's gesture for the length of a radius.

They created this graph by essentially placing a transparency on the screen and tracing out Sandra's path as she rode the double Ferris wheel.

As they began to plan for the construction of the qualitatively correct height versus time graph, they demarcated radii of the edge wheels using their thumb and forefinger (see Figure 9.4).

Placing the thumb at one end of a line and the forefinger at the other end of a line to demarcate length (thumb-forefinger gesture) would become an increasingly used gesture as their construction continued. The students then marked on a piece of scratch paper the length of the radii of the circular wheel with an R. They intentionally ignored the space from the ground to where Sandra starts her ride and the space between the two edge wheels so that they would not need to take it into account when constructing their height versus time graph. All of these activities allowed them to mark off the vertical axis of the graph in terms of R, the length of the radii of the edge wheels.

Episode 1

Andrew and Oscar concluded that it would take the middle bar, which the students called the big wheel, two full rotations for Sandra to start and return to the same position. In the absence of numerical information about the amount of time that it took for the middle bar or the edge wheels to make an entire rotation, the students had to chose an alternate way of marking time. Hence, they decided to separate the rotation of the middle bar into one-quarter turns and to mark the horizontal axis of their graph from zero to two, where one indicated one full revolution of the middle bar and two indicated a second full revolution, bringing Sandra back to her starting position. The students labeled the horizontal axis as t for time and chose the amount of rotation of the middle bar as a proxy for units of time. After spending several minutes discussing how they might proceed, Oscar and Andrew begin by drawing a set of axes (see Figures 9.5–9.7) that have time t, measured in terms of rotations of the middle bar, as the horizontal axis and height, measured in length of the middle bar, as the vertical axis. Along the bottom of the graph, Oscar draws hash marks in units of ¼, and along the horizontal axis.

> **Oscar:** Right. On this [vertical axis], we call it R1, R2, where each interval is equal to a radius of a Ferris wheel. So we know that at time [*pause*] at $t = 1$ (Figure 9.5), which is one complete revolution of the big wheel (Figure 9.6), 1, we're at 2R (Figure 9.7).

In Figure 9.5, Oscar marks on the horizontal axis the location of $t = 1$. He then moves his pen upward on the paper and gestures with his left hand a rotation of the wheel. Finally he points to the coordinate on the graph that corresponds to (1, 2R), where 1 is a rotation of the middle bar and 2R represents the length of two middle bars. Oscar's use of the rotational gesture coordinates movement as it appears on the applet with the particulars of the set of axes that he is drawing.

Episode 1 Analysis

Several cultural tools structure Andrew and Oscar's activity as they initially work to create their height versus time graph. On the one hand, Andrew's experience with the applet and knowledge of Ferris wheels allows

Figure 9.5 Oscar: So we know that at time . . . *t*, $t = 1$

Figure 9.6 which is one complete revolution of the big wheel [Makes a circular motion with his right finger]

Figure 9.7 I, we're at 2R. [Points at a space above $t = 1$]

him to make predictions about what occurs when the wheels move and the patterns that a rider makes when moving upon it. But their activity is also structured mathematically. In the construction of the horizontal and vertical axes, Oscar displays a set of disciplinary or cultural understandings of what is typically expected of inscriptions such as a height versus time graph.

The use of quarter rotations breaks up the horizontal axis into units that are meaningful for work with the applet, as quarter turns of the middle bar are easily coordinated via sight and button pressing of the mouse. The use of R1, R2, R3, and R4 express the work that Andrew and Oscar had done with the vertical versus horizontal displacement graph from task 1 and how radii had become meaningful within that activity. Furthermore, the use of radii in relation to circles and working with circular motion is common. For example, Oscar first draws hash marks, which is a common practice when generating graphs, but initially the values for vertical axis (height) are not in terms of feet or any other common unit of measurement; instead they are in lengths of the middle bar. After inscribing the graph with R1, R2, R3, and R4, Oscar stated, "each interval is equal to a radius of a Ferris wheel," communicating how his graph is to be interpreted in light of their previous activity. Creation of these axes establishes the ground on which Oscar, Andrew, and the interviewer can work at parsing the movement of the double Ferris wheel into meaningful components of the current task at hand. In making that connection, Oscar is also connecting his past experience with the applet and the double Ferris wheel with the graph that he is creating.

Oscar recruits different gestural expressions when describing the makeup of the horizontal axis, which integrate the experience of the movement of the wheel and hash marks on the horizontal axis. When Oscar says, "one complete revolution of the big wheel," he gestures a circle, counterclockwise, with his right index finger (Figure 9.6). At the same time, he points with the pen in his left hand on the paper above the hash mark corresponding to $t = 1$. The movement of his right hand in conjunction with pointing with the left establishes the meaning of that particular point as both a location in space and an experience with the double Ferris wheel. The location of a hash mark for $t = 1$ marks the horizontal positioning of a coordinate on a standard set of coordinate axes. Oscar and Andrew's cultural experience with mathematical inscriptions informs the setup of the graph and how it is that these hash marks are interpreted on the graph. However, what the horizontal positioning means in terms of the phenomenon is also tied to the experience of the phenomenon, in this case, a rotation of the middle bar. Unlike the lengths of the middle bar, which can be shown as relevant objects of scrutiny while the double Ferris wheel is stopped, time is a dynamic variable, and the meaning of particular intervals along the horizontal axis needs to be imagined as unfolding. Oscar's gesture (the circular motion with the right finger) invokes the prior movement of the double

Ferris wheel with the applet. Oscar's deictic pointing with his right hand and his gesturing with his left hand demonstrate and construct the multiple meanings for the horizontal axis by expressing the connection between the experience of the movement of the double Ferris wheel and the culturally significant experience of graphs and their structure.

In this episode, we want to highlight the intersection between the acting body and the encultured body. The encultured body is expressed via the set of mathematical assumptions that Oscar and Andrew bring to the creation of the graph and their experiences with the double Ferris wheel applet. The creation of a pair of axes each representing particular values for the two variables in question, height and time, shows sensitivity to how these graphs are created. Furthermore, their use of R, 2R, and so on and quarter-fraction increments also demonstrate this sensitivity. That circles have radii and a great deal of mathematics can be accomplished from attention to radii makes them a convenient and powerful reference point in lieu of having actual distances. The students also demonstrate a historical and cultural understanding of the movement of the Ferris wheel. On the applet, one-quarter revolution of the middle bar is easy to see and allows the students to stop the applet at a certain time interval and then count the distance from the bottom of the double Ferris wheel to Sandra's position on the ride in terms of lengths of the middle bar.

Oscar demonstrates the connection between the inscription and his experience of the double Ferris wheel in his gesturing of the rotation of the bar. By combining his gesture for the rotation of the bar with his marking of a particular point on the graph, he simultaneously organizes this location as a dynamic moment in space-time inseparable from the phenomenon of the Ferris wheel. The cultural memory of a graph as a mathematical inscription allows for Andrew and Oscar to develop an aspect of the organizing structure necessary for creating a qualitatively correct graph of height versus time. For Andrew and Oscar, these kinds of inscriptions have scales for time and height, and inscribing points on the height and time graph means coordinating the two quantities in a single point. Their embodied organizing activity also requires a connection of the inscription to the original phenomenon. Here Oscar and Andrew's memory of the movement of the wheel infuses the vertical and horizontal axes with meaning that relates to the double Ferris wheel.

Framing their interaction in terms of the active body highlights how their gestures and bodily movements play a key role in bringing the inscription and the applet into their peripersonal space, helping to bridge the two domains of experience. The thumb and forefinger gestures used to mark radius lengths on the double Ferris wheel highlight a certain aspect of the edge wheels and the middle bar, specifically the radii, and what it is about these aspects that is important, specifically the length. The parsing of the Ferris wheel into lengths and, later, breaking the movement of the Ferris wheel into quarter increments transform the phenomenon into what Goodwin (1994) calls an object of scrutiny. The process of creating the axes and populating the graph with a single point defines the relevant quantities that the participants want to focus on, as well as how those quantities will be annotated on the inscription. The particular relationship between height and time on the inscription creates an organizational structure from which the students can relate to the applet and Sandra's ride.

Episode 1 begins to create a ground from which Andrew and Oscar can relate to the double Ferris wheel mathematically. The ground on which they are working is mathematically and imaginatively informed. Mathematically, the creation of hash marks and the marking of points based on those hash marks is a convention derived from the culture of mathematics via their experiences as mathematics students and engineers. The ground is also imaginative as it is related via gesture and talk to the applet and the imagined movement of the double Ferris wheel. Furthermore, Andrew and Oscar have both ridden Ferris wheels at amusement parks and fairs, which adds to their experience and allows them to relate to the applet and the inscription using their knowledge from outside of the interview room.

From the perspective of the active body, we can see how gesture, body positioning, and gaze organize for Oscar and Andrew their activity with both the inscription and the applet. Their mathematical understandings of the coordinate axes generate a mathematical viewpoint from which they can align the applet and the inscription. This is no simple act of translation. In translation, aspects of the phenomena are moved to the inscription in a one-to-one correspondence, but the inscription, the applet, and the double Ferris wheel are separate. As is apparent in Episode 1, the students do not separate applet, inscription, and double Ferris wheel, but instead they work to bring them together and organize them along mathematical lines. Their knowledge of graphs establishes the necessary aspects of the double Ferris wheel.

Episode 2

In this episode, Oscar and Andrew place the second point on their graph and conjecture about what the connecting line from the origin to the point $(\frac{1}{4}, 2+\sqrt{22}R)$ should look like. The episode takes place immediately following Episode 1 as Andrew and Oscar continue generating the graph. In the process of illustrating how the students' activity adds to the inscription, we highlight the role that the imaginary body and the acting body play in organizing their activity. After Oscar and Andrew completed the axes and generated a first point for the qualitatively correct graph, Oscar moved back to the applet to generate more points on the inscription and further analyze the activity of the applet with regard to elaborating their inscription.

> **Oscar:** And then we've said at ¼, we're at 2R (Figure 9.8) plus (Figure 9.9) an amount that's less than R, which is a little bit less than 3R (Figure 9.10).

To continue his work in generating the qualitatively correct height versus time graph, Oscar returns to the applet, turning his body so he can look at both the applet and his inscription. He highlights that at time ¼, Sandra is at height 2R. He then uses his fingers to highlight a second radius length that will be added to 2R. He states that this length is less than R (implying less than a full radius length), and he gesturally indicates that this length is less than R by moving his fingers inward in a pinching motion. Consequently, Sandra's new height at time ½ is less than 3R. He then moves his body back to the table and the inscription. As Oscar returns to the table, Andrew remarks on the manner in which Sandra's height is changing:

> **Andrew:** Now, so it's been increasing, like in an... we can't say in a linear fashion, but it's been increasing all that time.

Andrew's comment notes that her height is increasing, but it is not increasing in a linear fashion. This statement will become important as Oscar and Andrew fill in the curves between the points that they have noted on the applet.

Episode 2 Analysis

At the beginning of the episode, Oscar positioned his body between the applet and the paper and looked toward the paper on the table. He extended his arm toward the computer with his thumb and index finger

	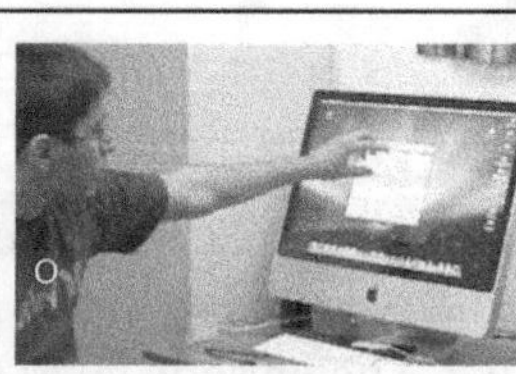	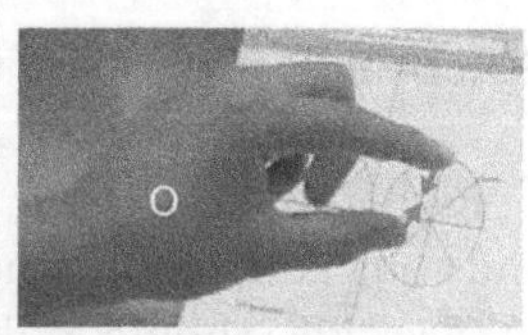
Figure 9.8 Oscar: And then we've said at ¼, we're at 2R [he then turns his head towards the screen]	**Figure 9.9** (pause) plus	**Figure 9.10** an amount that's less than R, which is a little bit less than 3R [he moves his body back to the table]

outstretched (Figure 9.8), denoting particular length segments, and stated, "And then we've said at ¼, we're at 2R." By positioning himself between the graph and the applet, Oscar brought both objects into his peripersonal space, coordinating them and allowing him access to both domains. His use of the prepositional phrase, "at 2R," indicates that Oscar is referencing the location 2R on the graph (rather than the height of 2R on the double Ferris wheel). The location 2R gets juxtaposed with the length 2R when Oscar placed his fingers over the different radii on the Ferris wheel and when he switched from saying "at" to saying "less than," which denotes measurement.

Unlike the first episode, where the quantities' heights and times were considered in isolation, to place a point on the graph, the quantities must be considered in tandem, as a location in space-time. What is necessary for the graph to be meaningful is a connection between the movement of the rider on the double Ferris wheel and what gets inscribed on the graph. By stopping the wheel, the students are able to concentrate on a particular moment in time. They then coordinate that moment in time with the height that the rider has reached. This allows the students to focus on the two highlighted quantities. The positioning of Oscar's body, his gaze, and his gesturing coordinates the point on the graph with the aspects on the applet that have been highlighted. Through these various bodily activities, Oscar further refines the record that he has generated of Sandra's height at various moments in time.

In this episode, we also highlight how the imaginary body coordinates with the active body. Andrew's utterance, "Now, so it's been increasing, like in an . . . we can't say in a linear fashion, but it's been increasing all that time," is not about a specific point in height-time. Rather, it references a specific aspect of the graph, and Andrew's use of this utterance implies the imaginary body in action. Andrew's utterance implies that he is imagining several activities. On the one hand, he imagines the movement of the applet as it moves in time. Second, he imagines the movement of an actual Ferris wheel, where the movement of the edge wheels and middle bar is circular, and there is an implicit comparison to either a graph of linear motion, which is a straight line, or actual linear motion, which does not move in a single direction. Third, Andrew imagines the graph that is about to be drawn as he determines how the curve between their sets of points will look as he and Oscar draw it on the paper. Andrew does not move his head from the applet, nor does he start the applet moving. Oscar's move back to the paper indicates to Andrew that he is considering where the next point is to be drawn corresponding to the Sandra's position at the moment when the middle bar has moved ¼ of a revolution. Andrew's statement in the past imperfect, "it's been," indicates that he is imagining the movement of the Ferris wheel in the past. However, simultaneously, he is also imagining the future construction of the graph, where there is a curve on the height versus time inscription that exists between the origin and the point that Oscar is about to draw. As Andrew and Oscar coordinate between the inscription and the movement of the rider on the double Ferris wheel as it is imagined and shown on the applet, the role of the imagination begins to expand. The inclusion of more points on the applet and the imagining of a path in space connecting the two moments in time $t = ¼$ and $t = ½$ necessitates a curve on the inscription connecting the two points. This curve is necessary by convention, but it also reflects Sandra's imagined movement through space as she continues her ride. The creation of a height versus time graph requires points and connections between points to become a graph and hence Andrew's direction to Oscar about what should be added to the inscription, that Sandra's height is increasing. Andrew also stated that it is not in a "linear fashion," indicating the manner in which the line that Oscar will draw should be formed. Andrew's utterance connects the future state of the imagined graph and his past experience of the movement of the Ferris wheel up until that particular moment.

The imaginary body does not operate only via Andrew's visualizing the movement of the double Ferris wheel from its starting point to the point at which the wheel has stopped. Rather, the imaginary body is operating throughout this episode by establishing for the students a set of expectations to guide their future activity. The imaginary body frames bodily experience as a constant moving into a somewhat preconceived future. The constraints of this future are given, in this context, by the students' disciplinary understandings of what is necessary on a graph of height versus time and their experiences with the movement of the applet and the double Ferris wheel. There is a set of necessary conditions for this graph, and they become the focal points for the students' continued activity. The focal points for the activity are the height that corresponds to a moment in time as expressed as a point on the height versus time graph and the behavior of the curve (linear, quadratic, sinusoidal, etc.) between the points based on the movement of the applet.

Andrew's utterance exemplifies the ways that the imaginary body works to create an inscription and develop mathematical meaning for Sandra's ride on the double Ferris wheel. His statement that "it's increasing" raises the question of what "it" he is referring to. First, he could be referring to Sandra's actual height as she rides the double Ferris wheel, suggesting that he is imagining an actual amusement park attraction and how an individual would be moving through space as the ride continued to move. In this case, Sandra, the imagined rider, is going higher and higher as the middle bar goes from ¼ rotation to ½ rotation. Second, he could be imagining the trace of the point on the applet that represents Sandra's position as the middle bar on the applet moves from ¼ rotation to ½ rotation. Finally, he could be imagining a proposed curve to be drawn on the inscription. From the perspective of the imagined body, Andrew actually refers to all three simultaneously. His utterance communicates the imagined juxtaposition of all three imaginary scenarios. This is one of the great powers of the imagination: the ability to bring together seemingly disjoint events and combine them into a cohesive and meaningful whole. The movement of the rider on the amusement park attraction, the trace of a point through space as the applet moves, and the curve that joints point at $t = ¼$ and $t = ½$ are all combined into a cohesive "it" that can then be characterized as "increasing" in a "nonlinear" fashion.

If we were to characterize Andrew's act here as one of translation, the interpretation would be very different. On the applet, as Sandra's point moves from ¼ rotation of the middle bar to ½ rotation, Andrew would identify what is happening to the height of the rider. Noticing that she was moving higher on the computer screen as the applet continued, Andrew could then characterize Sandra's height as increasing on the applet. Once they had returned to the graph of the height versus time, Andrew would remember his conclusion, and an increasing curve would be drawn. Andrew and Oscar may have to return to the graph to determine whether the graph is actually increasing in a linear or nonlinear fashion, and once they had decided, they could then draw a curve on the paper. But this is not what happened. Instead, immediately after stating that "it" is increasing, Andrew says that "it" is increasing in a nonlinear fashion, implying that the trace of the point representing Sandra's movement and the curve that will be drawn on the paper are simultaneously present. It should be noted that neither is actually present as a drawn aspect of the graph, as a trace on the computer or the amusement park ride; instead they exist only in Andrew's imagination.

Episode 3

In the third episode, Oscar placed the first two points on the graph, and then the two students located a third point (Figure 9.11). However, in between the points when $t = ¼$ and $t = ½$, Sandra's ride on the double Ferris wheel reaches its peak. Thus, in the process of finding the next point of interest to them, the two students seek to find where on the height versus time graph the maximum should be. In accomplishing this, they coordinate the movement of the applet and the plotting of particular points on the graph. In this episode, we illustrate the role that the expressive and imaginary bodies play in generating the inscription.

> **Oscar:** So 2R plus square root of 2 over 2R. So we keep it, it will reach a maximum in this next quarter, I'm thinking. So

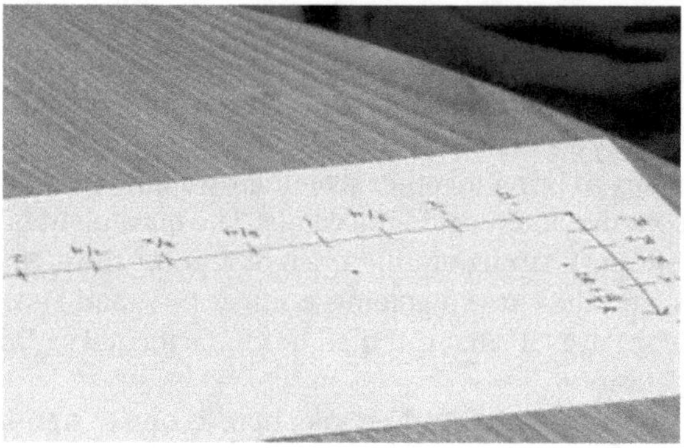

Figure 9.11 The height and time axes with two points inscribed upon them.

there... it's still... that's where it starts going down, so we have to take note of that.

Andrew: You reached the maximum, somewhere . . .

Oscar: Yeah, it reached the maximum somewhere in between. [Interviewer: Okay.] At this point, it's at what Andrew... 3R?

Andrew: Yeah, look, R (Figure 9.12) plus 2R, it's at 3R (Figure 9.13).

Oscar: So it's slightly... (Figure 9.14) Here it's at 3R and it does reach a maximum there... a maximum height. We could try to approximate that maximum height (Figure 9.15), but it's not going to be greater than R4 (Figure 9.16). So it must be in between R3 and R4. So, we'll just approximate (Figure 9.17).

Andrew: Somewhere around there. [marks the paper with a point]

Oscar: After that, we reach that R3.

Episode 3 Analysis

The imaginary body plays a strong role in organizing Oscar and Andrew's activity with the double Ferris wheel. The statement, "It will reach a maximum in this next quarter, I'm thinking," establishes an expectation for what is to occur and frames what the two students will look for as they run the applet. The maximum that he mentioned was referred to in previous conversations as being an important point of interest in the graph's construction. The use of the future simple "will reach" indicates the anticipation of a future state. The inclusion of the qualifier "I'm thinking" indicates some level of uncertainty. Furthermore, the preposition "in this next quarter" indicates when the expected maximum will occur—that it should happen sometime after he starts the applet moving and before the middle wheel reaches half of a rotation. In examining the next step in the construction of

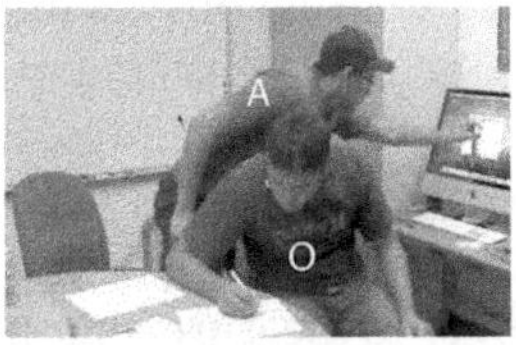

Figure 9.12 Andrew: Yeah, look R

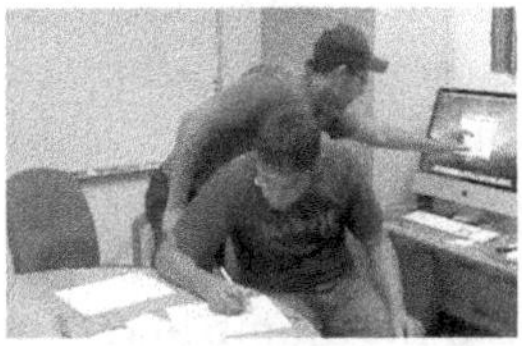

Figure 9.13 (opens his thumb and forefinger wider) plus 2R, it's at 3R

Figure 9.14 Oscar: So it's slightly (pauses)

Figure 9.15 Oscar: We could try to approximate that maximum height,

Figure 9.16 but it's not going to be greater than R4.

Figure 9.17 So it must be in between R3 and R4. So, we'll just approximate.

the graph, there is an expectation that Sandra's height with respect to the ground will decrease, and Sandra's motion will progress downward as the double Ferris wheel and the applet continue to move. Oscar also expects that that downward movement will begin somewhere in the next period of the movement of the applet. Unlike the previous episode, in which the students were coordinating between an unmoving applet and the inscription, now the students are dealing with the applet as it moves. The running of the applet, Oscar's pausing, and his final statement, "So we have to make note of that," all point toward the testing and fulfillment of their expectation that Sandra will reach a maximum height in that time interval.

The imaginary body sets forth a set of expectations that alter the kinds of activities in which Oscar and Andrew engage. Instead of being a specific height coordinated with a specific time, Oscar inscribed a point with a precise location on the height versus time graph, expressing a set of relative locations. Andrew's statement, "You reached the maximum somewhere . . . ?" and Oscar's confirmation, "Yeah, it reached the maximum somewhere in between," establish a relative location, in terms of time, for that point.

When the students inscribed this third point on the graph, they expressed the relativeness of the height in relation to other heights. Deciding where to locate this point is a complex activity in which Andrew and Oscar utilize their bodies, imaginations, speech, and coordinated action. While some of their actions are for themselves and intended to bring objects into peripersonal space or align them with the applet or the paper, other activities provided common frames of reference and background so that Oscar, Andrew, and the interviewer can communicate with each other. As Oscar moved back to the inscription, he asked Andrew, "At this point, it's at what Andrew... 3R?" Andrew used his thumb and forefinger to measure lengths on the applet while the bar and wheel are stopped, and then he added two lengths by moving the thumb and forefinger over different places on the screen. He then expressed the result, "It's at 3R," which Oscar then immediately marked on the inscription. Oscar then drew the point on the graph. Unlike the previous episode where Oscar coordinated most of the organizational activity with the inscription and the applet, Andrew takes a more active role in this episode. He carried out the activities that Oscar did in the previous episode, while Oscar faced away from the screen. In previous episodes, Oscar moved his body and head back and forth from the applet to the inscription to mark points, measure lengths, and determine what he would place on the inscription. In this episode, Andrew performed many of the activities that Oscar did previously, and their communication with each other eliminates the necessity for Oscar's head and body movement from the applet to the inscription.

An explanation for this is that, although Andrew played a less active role in the previous episode as they found the first point, Oscar's communication with Andrew allowed for Andrew to follow Oscar's bodily engagement, which included speech, and hence develop a shared repertoire for how to inscribe a point onto the graph in this particular situation. Oscar's question and his movement toward the paper initiated the set of bodily activities for Andrew. We see that this is an instance of the expressive body at work. Past experience, speech, and bodily movement interact to communicate and coordinate activities between people.

This activity sets forth for the students a set of boundaries on the height similar to those established for the time. They have the two heights for points at ¼ and ½, which describe a sort of boundary for "somewhere in between." Notably, when the two return to paper, it is not Oscar who begins to draw, but instead it is Andrew who annotated the maximum. Andrew marked a point on the paper that corresponded to an approximate maximum. At the exact same time, Andrew expressed how they arrived at this point. At this moment, both students' gazes are fixed on the paper. The coordination of gaze on the part of the two students, along with Andrew's annotation, indicates that Oscar was referring to the maximum on the inscription. Oscar stated, "it's not going to be greater than R4," which is the maximum possible height for any rider at any time on the double Ferris wheel, a fact that they established earlier in the session. Then Oscar stated, "so it must be between R3 and R4," indicating that the height must be greater than the point found when the middle bar has completed one half of a rotation.

This episode illustrates well a relationship between the expressive body and the imaginary body. In the course of structuring the inscription, students established a set of expectations for what they would be noticing in their activity with the applet, especially the maximum. However, in establishing the possibility of the maximum and then testing their conjecture against the movement of the applet, the students established a set of bounds, in terms of time, that constrained the maximum. In order to find the bounds that constrained the height of the Ferris wheel, Andrew and Oscar coordinated their bodily activity using speech and actions (marking the graph with thumb and forefinger, establishing R as a length on the applet and a hash mark on the coordinate axis, etc.). The imaginary body coordinates what both participants were paying attention to. But in establishing the specifics for where the point would be placed, the students' shared experience and speech played a crucial role in coordinating their actions.

Mathematically sophisticated individuals, such as Andrew and Oscar, have a collection of disciplinary and culturally significant experiences that allow them to talk about the phenomena in mathematical ways. However, the way in which their experience of the phenomenon and their mathematical experience coalesce into a graph is emergent. Because it emerges through their shared activities, this affords the two students new opportunities for action. In this episode, Andrew fully shared in the activity as he measured lengths on the applet, providing Oscar with information that in earlier episodes Oscar had arrived at by himself. In addition, Oscar does not need to look up at Andrew when Andrew asks him to "Look." Instead, the shared experience of creating the inscription and working with the applet and double Ferris wheel allows the two students to work together in new ways and maintain that the interpretations and constructions that they are making are in fact valid for both of them.

Episode 4

The final episode illustrates how the two students complete their construction of the graph from $t = 0$ to $t = 2$. They have already drawn the graph from $t = 0$ to $t = 1$ (Figure 9.18), and there has been a marked increase in the coordination of Andrew and Oscar's activities evidenced by the speed that they are working at and the amount of interdependence both display in their coordinated activities. The two students conjectured that the graph from $t = 1$ to $t = 2$ will be a reflection of the first part of the graph onto the second part of the graph with the center line at $t = 1$. In this episode, we

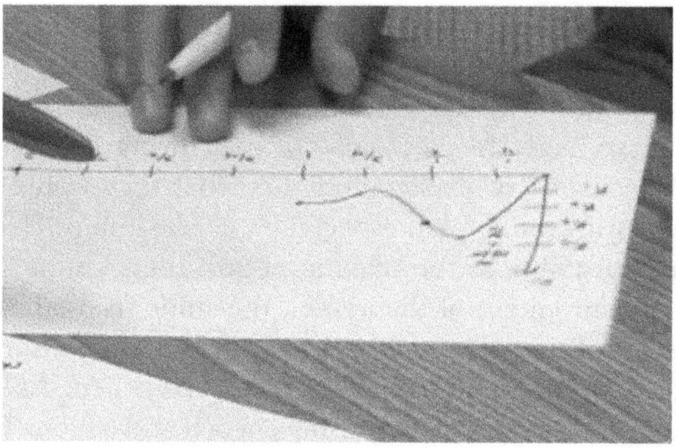

Figure 9.18 The graph of the height versus time from $t = 0$ to $t = 1$.

illustrate how they first posed this conjecture, tested the conjecture, and then constructed the graph.

> **Oscar:** And then... (Figure 9.19) we could... (Figure 9.20) 2 times. They should be the same (Figure 9.21).
>
> **Andrew:** Yeah, it seems like it. [Oscar begins to run the applet] Why not? It's some type of sinusoidal (Figure 9.22).
>
> **Oscar:** It starts decreasing, decreasing, decreasing (Figure 9.23). Yeah, 2R minus 2 over the square root of 2R (Figure 9.24). So if we go here (Figure 25), you have the same thing with the... (Figure 9.26)

Episode 4 Analysis

Oscar's gesture from one hand to the next hand (Figures 9.19 and 9.20) indicates that some aspect of the graph will be transferred from one side of the graph to the other. His pause and then "times 2" (Figure 9.21) indicates that some aspect of the graph should be doubled. It is apparent from their past interactions with each other and the applet that Oscar is indicating that there should be a reflection of one side of the graph over to the other. As in past instances, Oscar moved his body toward the applet as he spoke in the future tense, using "should." However, in this case, "should" has a weaker connotation of expectation than "will," which indicates a measure of uncertainty. His movement toward the screen indicates the desire to test his expectation that "they should be the same." Andrew continued to look at the paper, evidencing greater consideration of the graph. Andrew's utterance, "Why not? It's some type of sinusoidal," suggests a consideration of the shape of the graph. In this case, the students are trying to set forward an expectation about what the shape for the second half should look like. In Oscar's case, it is difficult to tell why he believes that the two sides of the

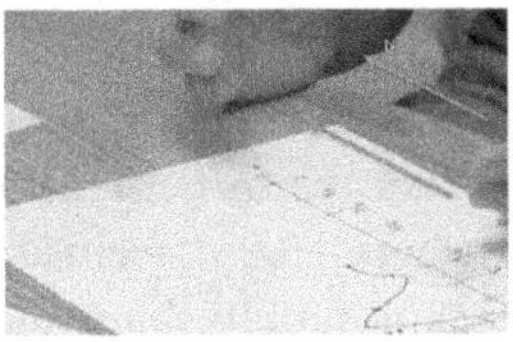

Figure 9.19 And then (pause)

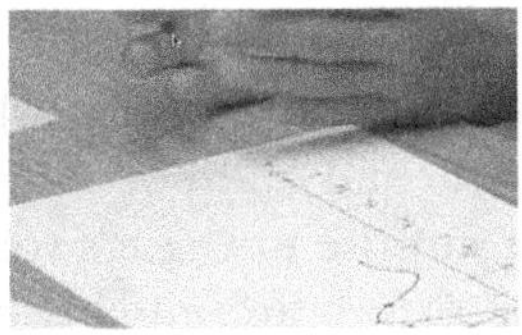

Figure 9.20 we could (pause)

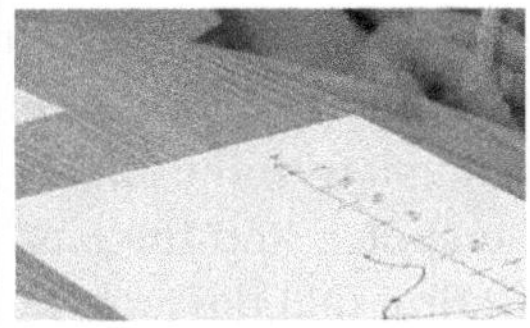

Figure 9.21 times 2 (pause) [turns his body towards the computer]. So you think (pause) they should be the same.

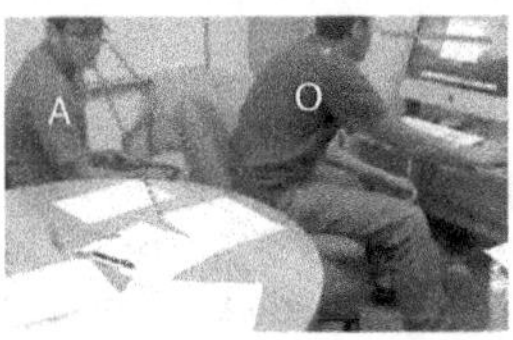

Figure 9.22 Andrew: Why not, it's some type of sinusoidal? [Oscar starts the applet.]

Figure 9.23 Oscar: It starts decreasing, decreasing... decreasing

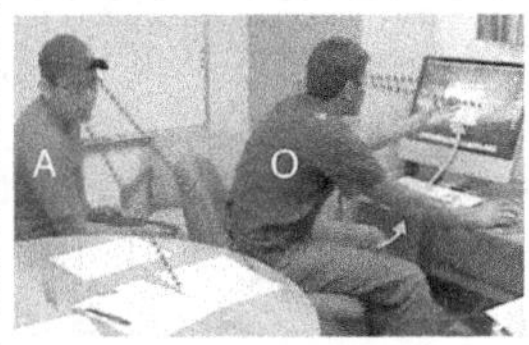

Figure 9.24 [Oscar stops the applet] Yeah, 2R minus 2 over the square root of 2R.

Figure 9.25 Oscar: So if we go here...

Figure 9.26 Andrew: you have the same thing with the [Andrew traces a line over the paper with his capped pen]

Figure 9.27 [Oscar draws a line over the place where Andrew has traced.]

graph should look the same. One possible explanation is that earlier the students had created a vertical and horizontal displacement graph in which the two sides of the vertical versus horizontal displacement graph were symmetric. Another explanation is that he had noticed that the vertical motion of the rider on the wheel seemed to repeat after one revolution of the big middle bar. Likely it is some combination of these two possibilities. In contrast, Andrew states, "Why not, it's some type of sinusoidal?" This could also be Oscar's unstated reasoning as well. But Andrew's statement indicates a relationship between an imaginary state in the future, the reflection of the graph from $t = 0$ to $t = 1$ over the line $t = 1$, and an encultured previous experience of sinusoidal graphs being symmetric.

Neither student is certain of his conjecture, which is evidenced by their testing it with the applet. Oscar and Andrew both consider the movement of the applet, and the lack of surprise in Oscar's voice (Figure 9.23) seems to confirm that the movement of the graph and his expectation of its shape did in fact match. Finally, Oscar stops the applet as he raises his hand toward the screen and says, "Yeah, 2R minus 2 over the square root of 2R" (Figure 9.24), affirming that the applet has reached the same location at the point when $t = 3/4$. The move away from the applet toward the paper to begin drawing the graph evidences that a certain agreement has been achieved. The fact that the two students never return to the applet as they finish the graph indicates that they agree that the graph is symmetric over the line $t = 1$.

As the two continue, Oscar includes the point that corresponds to the final stopping point of the applet. Andrew's statement and gesture, "you have the same with the..." (Figure 9.26), brings to Oscar's attention the shape of the graph as well as the point. Finally, Oscar traces over Andrew's undrawn line, finishing the drawing of the graph. The two students then work quickly to construct the graph, with Oscar drawing the points that correspond to quarter increments while Andrew draws the maximum and relative minimum. Unlike their previous activity, which was much slower and required the students to attend to the inscription and the applet simultaneously, this process seemed to encapsulate an entire set of assumptions about the nature of the inscription, the relationship between the inscription and the applet, and the validity of the activity that they used to create the inscription to this point.

The initial conjecture that the graph is symmetric over the line $t = 1$ is not specifically a conjecture about the applet or the movement of the double Ferris wheel but instead reflects a conjecture about the nature of the graph as a stand-alone object. The graph certainly relates meaningfully to the applet and the movement of the double Ferris wheel, but the conjecture does not directly reference them. As the students began to develop the graph, new opportunities for action became available that would not have been present otherwise, in this case, conjecturing about the two sides of the graph being symmetric. Testing the conjecture against the applet and confirming the conjecture set in motion a whole new set of activities. Once their reasoning has been tested against the applet, the students can safely go about using their new knowledge to finish the inscription. The finishing of the inscription played out in activity that was similar to how the students first created the graph from $t = 0$ to $t = 1$. However, this time the shared repertoire of experience and the established set of modal engagements that they used previously made the creation of the second half of the graph a quick process.

CONCLUSION

In this chapter, we took an embodied cognition perspective on modeling. From this point of view, modeling is more completely understood as an embodied organizing activity that engages the body, cultural tools, and imagination. For this primary position, we set forth three characteristic features that highlight modeling as an embodied organizing activity. The first is that bodily engagement organizes a particular experience by aligning seemingly different perceptual fields in space-time, facilitating the perception of that experience as mathematical in nature. For example, we demonstrated how movement of the body between applet and the inscription brought the experience of the applet and the inscription together. Oscar and Andrew's use of the thumb and forefinger gesture, in conjunction with speech and pointing, coordinated the lengths on the applet with the hash marks on the inscription. The students' bodily engagement also coordinated the imagined movement of an actual double Ferris wheel with the movement of the applet and the inscription. Oscar's circular gesture to represent time referenced the actual experience of moving in a circle on the double Ferris wheel as a dynamic experience that occurs over time. And the students' brought their experience as mathematically savvy individuals to their design of the inscription and their understanding of the applet and the movement of a double Ferris wheel. The facilitation of the coordination of these different realms of experience occurred in imagination as the students juxtaposed these experiences against one another in ways that could or would not occur in actual physical space.

The second characteristic of modeling as an organizing activity is that the use of mathematical tools shapes our expectations, expanding and constraining how we perceive a particular experience. This characteristic was demonstrated in the ways that Oscar and Andrew's work at constructing the graph shaped their interpretation of what was happening on the double Ferris wheel. The experience being shaped by Oscar and Andrew's mathematical activity was their experience of the applet and that of the double Ferris wheel. Andrew's mention of the linearity of Sandra's path as he looked at the applet is a clear example of how an experience can become mathematical. Furthermore, as the students' creation of the graph accelerated and they began to construct more of the height-versus-time graph, the role of symmetry and the periodicity of the graph also transformed their understanding of the double Ferris wheel and the applet. They saw Sandra's ride as symmetric about a certain point and her path at the beginning of the ride as mirrored at the end of the ride. Symmetry and periodicity became important aspects of their understanding of Sandra's ride as a whole.

The third characteristic of modeling as an embodied organizing activity is that by organizing a particular experience as a mathematical experience, new perceptions are made possible, enhancing our understanding of the original experience and facilitating new mathematics. The realization that Sandra's path on the double Ferris wheel is symmetric transformed Andrew and Oscar's construction of the height-versus-time graph. Once they completed one half of the graph, they drew a mirror image of the second half of the graph. However, Andrew and Oscar could not have taken full advantage of their understanding of the symmetry of Sandra's ride on the double Ferris wheel without their previous mathematical activity. One half of the height-versus-time graph needed to be constructed, and Sandra's path on the applet and on the double Ferris wheel had to be coordinated with how her path would be represented on the inscription.

In this chapter, we have characterized modeling as an embodied organizing activity rather than an act of translation (Blum & Niss, 1991). In acts of translation, the realms of experience, in this case the applet and the inscription, are fully formed, one as mathematical and the other as experiential, and what is needed is to form connections between those realms of experience. In the mathematical realm of the inscription, abstract conventions govern the creation of graphs, the setting of axes, and the interpretation of graphs once they are created. In the world of the applet, the edge wheels and the middle bars move around, and Sandra has a particular path in which she moves up and down. If we view modeling as translation, the process of modeling creates the relationships between these two worlds, but they never really impede on each other. When viewing modeling as an embodied organizing activity, these seemingly separate realms of experience transgress on one another, and a new realm of experience, the model, is generated. Because the creation of new mathematics leads to new mathematical understandings, which in turn generate new opportunities for action, which then lead to new mathematics, the act of modeling never fully ceases until the individual or groups engaged in the action move away from the model and onto other activities.

From the perspective of translation, the model is the inscription or the function derived to represent the phenomenon. But viewing modeling as an embodied organizing activity, the model is never fully complete, and hence the model is under continual construction and development. In the process of creating the height-versus-time graph, Andrew and Oscar developed a record of their embodied organizing activity. The inscription is one record of their activity, what Latour (1987) calls an immutable mobile that can be passed to others and hence understood in a variety of ways as new users engage with the inscription. However, the inscription is not the only record of their embodied organizing activity. Their body movements as they constructed the graph, their talk, and the mental actions of their imagination also generate a record, a bodily record, of this embodied organizing activity. This record is kept in muscle memory, in imagination, and in the interplay between the two students.

Analytically, we examined this embodied organizing activity from four different perspectives: the active body, the imaginary body, the expressive body, and the encultured body (Hall & Nemirovsky, 2011). For the purposes of our discussion, these four framings might appear as separated, but when engaged in activity, there is no separation. Instead, all of the bodies' faculties, expressive, imaginative, active, and cultural, are recruited. Modeling as translation focuses on the inscription, the physical double Ferris wheel, and on the applet and the connections made among these three separate realms of experience. However, it assumes Andrew and Oscar understanding of the inscription can be extricated from their experience of its creation and the subsequent activity carried out with that model. We illustrated how it is that modeling as an embodied organizing activity recruits the entirety of the body and the students' disciplinary knowledge and even their knowledge of the double Ferris wheel. We saw Oscar bringing both the graph and the applet into his peripersonal space via aligning his body, positioning his gaze, and reaching out to point at the screen while speaking to Andrew and the interviewer. These activities organized his work in meaningful ways and set forward future courses of action for him and Andrew.

In Episode 1, Oscar drew the axes for the height versus time graph and explained how the axes related to the movement of the double Ferris wheel. The students' understanding of the relationship between height and time connect to their prior mathematical experience with such graphs. Andrew and Oscar demonstrated sensitivity to the need for coordinate axes and realized that the coordinate axes need specific quantities associated with them and that points are inscribed in relation to those axes. Furthermore in Episode 3, Andrew and Oscar's conjecture that the maximum lies "somewhere between" $t = 1/4$ and $t = 1/2$ highlights how meaning emerged as they engaged in model construction. When they returned to the inscription, their perception was geared toward affirming the conjecture and establishing a relative height related to a relative time. This same sort of conjecture testing is also exemplified in testing whether the graph of the height versus time was symmetric.

In each of these cases, enculturation and imagination worked together to shape the makeup of the graph and the perception of Sandra's ride on the double Ferris wheel. The mathematical tools that both Andrew and Oscar brought to bear established a field of perceptuomotor activities that shaped how they interacted with the double Ferris wheel applet. New activities arose with the expectations that they had for what the graph should look like and the possible relationships between the applet and the inscription. It is like writing a shopping list to go to the grocery store. Given a particular goal, buying groceries for the week, we create a set of plans that highlights what is important at the grocery store for us. Upon entering the grocery store, the list guides what we notice and what is of greatest importance. Other aspects of the store may become interesting and we may buy other things, but in the case of our particular goals, the list sets forward the path of our future action.

If our experience of the store is limited, we may have to spend considerable time rereading the list as we pass through different aisles in the store. But as we shop more or spend more time walking through the store, we begin to understand the particular structure of the store and can use our knowledge of the store to refine our shopping list. The next time we write a shopping list, instead of basing it solely on our needs at home, we can rewrite the list in terms of how the store is laid out. Thus, when we take our list into the store, it is written so that as we move through the store, we only need to move down the list. Similarly, as Oscar and Andrew worked increasingly with the double Ferris wheel to create the graph, they utilized their experience with the double Ferris wheel to shape how the graph is laid out. This is a possible explanation for the increase in speed and efficiency with which they worked after their initial forays connecting the double Ferris wheel and the graph. As they worked more and more on the task at hand, their perceptions became attuned to the particular connections that they wanted to make between the graph and the applet, and their work became more efficient.

As Andrew and Oscar created their graph, new opportunities for action became available. We saw this in their conjecture regarding the symmetry of the graph across the line $t = 1$. The conjecture of symmetry is mathematical in nature and so belongs to the field of mathematical experience. It was made available by their past experiences and then it was confirmed by testing their conjecture against the movement of the applet. In the course of modeling phenomenon, students are organizing the phenomena into mathematical experiences. In so doing, they align a phenomenon with their previous mathematical experiences and bring even more of their mathematical tools to bear in organizing their experience, creating the inscription, and understanding the phenomenon.

One of the challenges that teachers face is how to connect mathematics to students' everyday experiences. By engaging in the activity of modeling as an embodied organizing activity, students can create meaningful connections between their everyday experiences and the mathematics they learn in school. Providing opportunities for students to develop a record of their mathematical activity involves allowing them occasions to bring together their mathematical and everyday experience, use their experience of mathematics to enhance their understanding of the phenomena, and notice how certain qualities in the phenomena might be highlighted. We conjecture that such opportunities can enhance students' experience of the world and of mathematics, and develop in them productive dispositions toward mathematics.

REFERENCES

Alibali, M. W., & Goldin-Meadow, S. (1993). Gesture-speech mismatch and mechanisms of learning: What the hands reveal about a child's state of mind. *Cognitive Psychology, 25*, 468–468.

Alibali, M. W., Kita, S., & Young, A. J. (2000). Gesture and the process of speech production: We think, therefore we gesture. *Language and Cognitive Processes, 15*(6), 593–613.

Alibali, M. W., & Nathan, M. J. (2007). Teachers' gestures as a means of scaffolding students' understanding: Evidence from an early algebra lesson. In R. Goldman, R. Pea, B. Barron, & S. J. Derry (Eds.), *Video research in the learning sciences* (pp. 349–365). Mahwah, NJ: Erlbaum.

Arzarello, F., Paola, D., Robutti, O., & Sabena, C. (2009). Gestures as semiotic resources in the mathematics classroom. *Educational Studies in Mathematics, 70*(2), 97–109.

Blum, W., & Niss, M. (1991). Applied mathematical problem solving, modelling, applications, and links to other subjects–State, trends and issues in mathematics instruction. *Educational Studies in Mathematics, 22*(1), 37–68.

Coslett, H. B. (1998). Evidence for a disturbance of the body schema in neglect. *Brain and Cognition, 37*, 529–544.

Goldin-Meadow, S., Nusbaum, H., Kelly, S. D., & Wagner, S. (2001). Explaining math: Gesturing lightens the load. *Psychological Science, 12*(6), 516–522.

Goodwin, C. (1994). Professional vision. *American Anthropologist, 96*(3), 606–633.

Goodwin, C. (2000). Action and embodiment within situated human interaction. *Journal of Pragmatics, 32*, 1489–1522.

Goodwin, C. (2003). The semiotic body in its environment. In J. Coupland & R. Gwyn (Eds.), *Discourses of the body* (pp. 19–42). New York, NY: Palgrave Macmillan.

Gravemeijer, K. (1997). Solving word problems: A case of modelling? *Learning and Instruction, 7*(4), 389–397.

Hall, R., & Nemirovsky, R. (2011). *Modal engagements as the conceptual fabric of embodied mathematical cognition.* Manuscript in preparation.

Herscovics, N. (1996). The construction of conceptual schemes in mathematics. In L. Steffe, P. Nesher, P. Cobb, G. Goldin, & B. Greer (Eds.), *Theories of mathematical learning* (pp. 351–379). Mahwah, NJ: Erlbaum.

Holmes, N., & Spence, C. (2004). The body schema and the multisensory representation(s) of peripersonal space. *Cognitive Processing, 5*(2), 94–105.

Kennedy, D. P., Gläscher J., Tyszka J. M., & Adolphs, R. (2009). Personal space regulation by the human amygdala. *Nature Neuroscience, 12*(10), 1226–1227.

Key Curriculum Press. (2011). *Double Ferris wheel.* [Internet application software]. Available at http://www.keymath.com/x3361.xml

Ladavas, E., Di Pellegrino, G., Farne, A., & Zeloni, G. (1998). Neuropsychological evidence of an integrated visuotactile representation of peripersonal space in humans. *Journal of Cognitive Neuroscience, 10*, 581–589.

Latour, B. (1987). *Science in action.* Cambridge, MA: Harvard University Press.

Nemirovsky, R., & Ferrara, F. (2009). Mathematical imagination and embodied cognition. *Educational Studies in Mathematics, 70*(2), 159–174.

Nemirovsky, R., & Monk, S. (2000). "If you look at it the other way...." An exploration into the nature of symbolizing. In P. Cobb, E. Yackel, & K. McClain (Eds.), *Symbolizing and communicating in mathematics classrooms* (pp. 177–221). Mahwah, NJ: Erlbaum.

Nemirovsky, R., Rasmussen, C., Sweeney, G., & Wawro, M. (2011). When the classroom floor becomes the complex plane: Addition and multiplication as ways of bodily navigation. *Journal of the Learning Sciences, 21*(2), 287–323.

Previc, F. H. (1998). The neuropsychology of 3-D space. *Psychological Bulletin, 124,* 123–163.

Radford, L. (2009). Why do gestures matter? Sensuous cognition and the palpability of mathematical meanings. *Educational Studies in Mathematics, 70,* 111–126.

Rizzolatti, G., Fadiga, L., Fogassi, L., & Gallese, V. (1997). The space around us. *Science, 277,* 190–191.

Vaishnavi, S., Calhoun, J., & Chatterjee, A. (1999). Crossmodal and sensorimotor integration in tactile awareness. *Neurology, 53,* 1596–1598.

CHAPTER 10

MORE THAN MERE HANDWAVING

Gesture and Embodiment in Expert Mathematical Proof

Tyler Marghetis
University of California, San Diego

Laurie D. Edwards
Saint Mary's College of California

Rafael Núñez
University of California, San Diego

For the budding mathematician, the ability to comprehend and produce rigorous proofs marks the transition from a novice student or consumer of mathematics to a more expert practitioner. By the time a student begins graduate school in mathematics, he or she is capable of both understanding and creating formal proofs of known mathematical results and is expected to generate new proofs and theorems. Often the focus is on the proofs themselves, the stable, shared outcomes of a process of proving.

Emerging Perspectives on Gesture and Embodiment in Mathematics, pages 227–246

In this chapter, by contrast, the focus is on proof as an active practice carried out by knowledgeable mathematicians, looking specifically at the relationship among gesture, speech, thought, and technical definitions. This case study is part of a larger ongoing research project on expert proof, in which we focus on how advanced students of mathematics conceptualize proof, how they collaborate to produce a proof, and how they coordinate multiple modalities during their practice, modalities that include speech, gesture, and inscriptions, both symbolic and graphic.[1] Our general thesis is that these resources are deployed in a complex but coherent manner to construe and scaffold the building of the desired intellectual product, the mathematical proof.

The goals of this chapter are twofold. First, we argue for the importance of gesture in mathematical practice, including proof, and review recent research that places the body at the center of mathematical practice. Second, we report on a case study of mathematical proving that treats proof as a human activity, extended over time and performed in a social and material context (Edwards, 2010; Marghetis & Núñez, 2013). Specifically, we document the various kinds of gestures produced by mathematical experts while they elaborate a proof and argue that these spontaneous gestures suggest that experts' reasoning about mathematical concepts is fundamentally metaphorical and embodied. By examining the practice of proving from a multimodal perspective, we hope to shed light on the nature of proof itself, as a product but also as a human practice.

FROM PRODUCTS TO PRACTICE IN THE STUDY OF MATHEMATICS

Traditionally, foundationalist accounts of mathematics (e.g., Benacerraf & Putnam, 1983) have focused exclusively on the products of mathematical activity—proofs and theorems—while ignoring the rich practices that contribute to, and potentially constitute, the core of mathematics. This tendency to analyze static, abstract products rather than dynamic, human practices is reminiscent of early formalist accounts of natural language, where, as Kendon (2008) notes, "what is transferred to paper is abstracted away from what is actually done within an enacted utterance" (p. 357).

Over the last few decades, researchers in mathematics education have broadened their focus beyond proof as a product to investigate proving as a cognitive *process* and a form of regimented discourse (e.g., Balacheff, 1991; Boero, 2007; Hanna, 1991; Knuth, 2002), although this literature has generally focused on secondary school mathematics (typically geometry), rather than on more advanced levels. Only recently, however, have these investigations attended to modalities beyond speech and inscriptions. But without a broader scope that includes the mathematician's body, essential aspects of communicative practice are invisible (Goldin-Meadow, 2003; Goodwin, 2000; Kendon, 2004, 2008; McNeill, 1992, 2000). Indeed, the particular form of kinesis represented by hand and arm gestures has unique communicative potential during situated proving. Manual gestures are often enacted unconsciously and can convey information that might be consistent with, complementary to, or different from the content of accompanying speech (Kita, 2000; McNeill, 1992) or inscriptions (Marghetis & Núñez, 2013). In fact, we would go further to propose that gesture and other bodily movement is essential not only in communicating about mathematics but also in the intellectual construction of mathematics. In short, the mathematician's body may be a constitutive part of his or her situated proving.

This claim requires empirical support, of course, and later in this chapter we describe a recent study that supplies such evidence. But before turning to the case study, we briefly review some of the recent literature on co-speech gesture and mathematics.

GESTURE: MEANINGFUL MOVEMENTS OF THE BODY

Gesture is universal, typically unmonitored by the speaker, and essential to communication (Goldin-Meadow, 2003; Kendon, 2004; McNeill, 1992). Most important, gesture offers a "window into the mind" (Goldin-Meadow, 2003, p. 1), in that gestures can provide clues to how a person is conceptualizing the situation being discussed. When co-produced with abstract thinking, a speaker's gestures may parallel the metaphorical reasoning they exhibit in speech (Cienki, 1998; McNeill, 1992, 2000; Núñez, 2006), and give us insight into their representation of mathematical concepts and solution strategies (e.g., Alibali, Bassok, Solomon, Syc, & Goldin-Meadow, 1999; Garber & Goldin-Meadow, 2002). These meaningful movements of the hands and body, moreover, are meaningful not only for the speaker but also for the listener: "perceiving hand movements during speech modulates the distributed pattern of neural activation involved in both biological motion perception and discourse comprehension, suggesting listeners attempt to find meaning, not only in the words speakers produce, but also in the hand movements that accompany speech" (Dick, Goldin-Meadow, Hasson, Skipper, & Small, 2009, p. 3509). And the influence of gesture is not limited to concrete, literal language: Evidence from electrophysiology suggests that the cross-modal coordination of speech and gesture shapes the neural response to metaphorical language (Cornejo et al., 2009). Because speech and gesture differ in their communicative possibilities, their combination may "package" complementary forms of information within the same discourse: linear, symbolic verbal language, on the one hand, and global, instantaneous imagery, on the other (Goldin-Meadow, 2003; Kita, 2000; McNeill, 2000).

Various means of categorizing or characterizing gestures have been developed. In this study, an initial distinction is made between gestures whose purpose is to facilitate the interaction between the speaker and interlocutor and gestures that refer to the content of the conversation. Bavelas and her colleagues have called these two types of gestures "interactive" and "topical," respectively (Bavelas, Chovil, Lawrie, & Wade, 1992). The analysis elaborated here focuses on these "topical" (also called "representational") gestures, especially the way they can embody abstract mathematical concepts during proving. McNeill (2000) has further characterized gestures along four dimensions: iconicity (resemblance to concrete referents), metaphoricity (reference to abstract entities), deixis (pointing or context-dependence), and "temporal highlighting" or beats (giving emphasis through repetition, i.e., producing "beats" in time with speech). Kendon (2004) makes even further distinctions, discussing three varieties of gestural representation that are distinguished by the relationship of the gesture to its referent: enactment, depiction, and modeling. In enactment, the motor action is meant to reproduce some features of the activity being represented. For example, gesturing in the air as if you were drawing a graph would be an enactment of the physical act of creating a graph. In depiction, the gesture "creates an object in the air" (p. 160), typified by using the index finger to trace an object's shape. In the example of graphing, while enactment would represent the action, a finger tracing a graph in the air would be a representation of the finished product (in this case, the act of drawing and the actual graph that is produced may be manifested in similar gestures). In modeling, the gesturing body part stands in for another object, as when a fist represents a stone, or a speaker indicates a tangent line by holding his or her palm at an angle. As we shall see, these distinctions will be useful for describing and interpreting the gestures generated by doctoral students while they work together to find a proof.

MATHEMATICS AND GESTURE

Gesture is prevalent in mathematical discourse as in other kinds of speech (McNeill, 1992). Previous research on gesture and mathematics has looked at various mathematical tasks, from simple (e.g., learning to count; Alibali & diRusso, 1999) to advanced (e.g., discussing differential equations; Rasmussen, Stephan, & Allen, 2004). Gesture can offer insight into how students think while solving problems and into the effectiveness of teachers' communication (Alibali et al., 1999; Edwards, 2009; Garber & Goldin-Meadow, 2002; Goldin-Meadow, Cook, & Mitchell, 2009; Goldin-Meadow,

Kim, & Singer, 1999; Goldin-Meadow, Nusbaum, Kelly, & Wagner, 2001; Goldin-Meadow & Wagner, 2005). In a study of gestures used during algebra problem solving (Alibali et al., 1999), for instance, researchers found a relationship between the form of representational gestures and the semantic content of the mathematics problem. If a problem was stated in a way that involved step-by-step, discrete changes, the gestures that participants used to describe that problem also tended to be "discrete" (e.g., a sequence of taps or beats). Conversely, a problem phrased in terms of continuous change elicited sweeping, arcing, or other "continuous" gestures. Gesture type also tended to reflect the particular kind of solution attempted (one type representing a view of the problem as continuous and the other as discrete). However, when the participant's speech and gesture did not match, they tended to use the strategy associated with their gestures rather than that associated with their speech. As the authors noted, "Our findings suggest that mental representations often include visual or perceptual information, which may at times be more readily expressed in gesture than in speech. In these cases, spontaneous gestures can be a valuable tool for illuminating mental representations" (Alibali et al., 1999, p. 332).

While early investigations into mathematics as an embodied activity generally focused on mathematical language (Fauconnier & Turner, 2002; Lakoff & Núñez, 2000; Núñez, 2006), typically taken from textbooks and conventional language, more recently mathematical gesture has provided converging evidence for the role of embodied, possibly unconscious, cognitive processes like conceptual metaphor (Lakoff & Johnson, 1980) and fictive motion (Talmy, 2000). Spontaneous gesture has provided evidence that arithmetic, for instance, is conceptualized using metaphors of Object Collection or Motion Along a Path (Lakoff & Núñez, 2000; Núñez & Marghetis, to appear). Analyses of gestures for fractions, geometry, and graphing have revealed that implicit, embodied conceptual mappings are manifested in bodily action (e.g., Edwards, 2009; Font, Bolite, & Acevedo, 2010). Even mathematicians' gestures about highly technical mathematical concepts reveal such embodied thought, at least when produced in a pedagogical context (Núñez, 2006, 2008).

Conclusions drawn from classroom contexts and mathematically naïve undergraduates, however, may not extend to the non-pedagogical activities of expert mathematicians. Here we elaborate on a study that utilized video and audio data recorded during sessions in which doctoral students in mathematics worked together to create a mathematical proof for a conjecture that was new to them. In Marghetis and Núñez (2013), we applied the tools of gesture studies, cognitive linguistics, and embodied cognition to examine the spontaneous co-speech gestures of doctoral students as a means of better understanding their thinking about a proof-in-progress (see also Edwards, 2010). In contrast to many studies of gesture and mathematics, this setting was not pedagogical. Instead, the participants in this study were asked to do something that they had done many times before—create a valid proof—but because the conjecture was new to them, this task nevertheless represented an intellectual challenge and offered a context for sophisticated mathematical discourse. This approach offers an opportunity to examine proving as a central practice of mathematicians. The participants' gestures thus accompanied authentic mathematical practice and have the potential to reveal textured aspects of the participants' mathematical thought.

A CASE STUDY OF EXPERT PROOF

When mathematicians work on a proof, they talk, scribble, on blackboards and in notebooks; stare pensively into empty space; and, crucially, gesture even when working alone. These co-thought and co-speech gestures might be mere "handwaving," empty accompaniments to "core" processes of logical inference. As we have seen, however, there is mounting evidence that gestures are central to, and revealing of, abstract thought more generally. Might expert mathematical discourse reveal a relationship between gesture form and the inferential structure of the mathematical argument?

To answer this question, we went to a natural site of mathematical proving: the academic office. We arranged for twelve doctoral mathematics students from a large American research university to collaborate on a non-trivial proof, working in pairs at a blackboard. They had been enrolled in the graduate program for various lengths of time, from ten months to three years, and were working in a variety of subfields within mathematics. During a 90-minute session, the participants were first interviewed about their mathematical specializations, as well as their approaches to learning and teaching proof. After the interview, the pairs of students were asked to prove the conjecture below, provided to them on a sheet of paper:

> Let f be a strictly increasing function from [0, 1] to [0, 1].
> Prove that there exists a number a in the interval [0, 1]
> such that $f(a) = a$.

This problem was selected, in part, because the mathematician who proposed it reported that when proving the theorem, he experienced a palpable sense of motion. We also checked with faculty at the university to determine that the theorem was not presented in any of the graduate analysis courses, and thus was likely to be new to the students. Although the problem comes from real analysis, it does not require specialized or advanced knowledge within that subdiscipline. The proving of this theorem, therefore, presented an excellent opportunity to challenge and examine the mathematical thinking of the participants.

Participants were given up to 40 minutes to solve the problem, working in a room with a blackboard, on which they were asked to record their work. The entire session was videotaped, but the investigator (the second author) was not present while the students worked on the proof. Once the pair was satisfied with the proof, or the 40 minutes had passed, the investigator returned to the room, and the participants were asked to explain their proof or their progress to that point.

Here we focus on an analysis by Marghetis and Núñez (2013) of the mathematical speech and gestures from these video-recorded sessions. The analysis concentrated on gestures used to represent mathematical content, that is, gestures that were co-produced with talk of particular mathematical concepts. Specifically, Marghetis and Núñez examined representational gestures produced in conjunction with two categories of conceptual content: static and dynamic mathematical ideas.

By static ideas, we mean those mathematical concepts that are hypothesized to rely on static spatial schemas such as containment—ideas like mathematical sets, which can be conceptualized metaphorically as containers for their elements. By contrast, dynamic ideas are mathematical concepts that are thought to rely for their conceptualization on intuitions of space and motion. This distinction between static and dynamic ideas grew out of work by Talmy (2000), who noted that language typically used to express motion is often used in the absence of literal motion, a phenomenon known as fictive motion. For instance, we say that "the road runs through the hills of Tuscany," even though roads are static and thus incapable of literally running anywhere. Lakoff and Núñez (2000) noted that a similar phenomenon exists for technical mathematical concepts—including the notions of functions, continuity, and limit—where the technical definitions are entirely static and yet mathematicians reliably talk about such concepts using dynamic language.

For illustration, consider the concept of a limit, a central notion in calculus. Technically, the limit of a function is defined by a chain of inequalities:

Let a function f be defined on an open interval containing a,
except possibly at a itself, and let L be a real number.
Then $\lim(x \to a)\ f(x) = L$ means that, for all $\varepsilon > 0$, there exists $\delta > 0$,
such that whenever $0 < |x - a| < \delta$, then $|f(x) - L| < \varepsilon$.

Note that the limit notation includes a small arrow, which might suggest that the definition of a limit of a function would include some notion of dynamism. The technical definition of a limit, however, refers only to static universal and existential quantifiers, static numbers, motionless arithmetic differences, and static inequalities. Nowhere in this definition is there any mention of movement. Mathematicians, in contrast, speak of a function "tending to," "moving toward," or "reaching" a limit—all of which, unlike the formal definition, invoke a sense of motion (Núñez, 2006). These expressions are all instances of fictive motion, invoking a trajector (moving actor) in motion across a landscape (non-moving background)—despite the fact that there is no literal motion involved. Similar instances of fictive motion inject dynamism into a wide range of statically defined mathematical entities. A function, for example, is formally defined as a static relation between two sets, the domain and the range, but mathematicians nevertheless routinely describe functions dynamically as "*reaching* an asymptote," "*going down* towards a minimum," or "*oscillating*," in each case evoking a construal in which an imaginary trajector travels along the path of the function (Núñez, 2006). Fictive motion is similarly at work when we say that sequences are "approaching," "decreasing," or "converging," and when arithmetic is construed as motion along a number line. Specific forms of dynamism, therefore, are present throughout the language of mathematics, showing up in the discourse surrounding continuity, functions, and even arithmetic, which lends credence to the claim that mathematical thought builds on dynamic, metaphorical, and embodied construals.

While previous analyses have argued that these ideas rely in varying ways on spatial intuitions like motion or containment (Lakoff & Núñez, 2000), this remained largely speculative: There was no empirical evidence that dynamic, spatial intuitions played a role in real-time, expert practice. Language, after all, can mislead, and the archaeology of language is a fraught enterprise: "*fast*ing," for instance, has nothing to do with speed or motion. It's entirely possible that the rampant fictive motion in conventional mathematical language is similarly misleading, suggesting dynamism where there is none. As we have pointed out, however, mathematical discourse is not limited to speech but also involves gesture. Can gestures, meaningful movements of the body, provide additional evidence for dynamism in mathematical thought?

EVIDENCE FOR FICTIVE MOTION IN GESTURE

To zoom in on this mathematical content of interest, Marghetis and Núñez (2013) generated a list of lexical items thought to be associated with fictive motion or dynamic conceptual metaphors, drawing on previous theoretical work within cognitive linguistics (Talmy, 2000). These included mathematical terms (e.g., function, continuity, limit, contain), verbs of motion (e.g., to cross, to move, to jump), and spatial terms (e.g., up, between, left).

Mathematical terms referring to static concepts (e.g., set, point) were also identified. Of course, many everyday concepts are open to multiple construals, and technical mathematical concepts are no different. The notion of continuity, for instance, can be construed dynamically as movement without jumps—but also as "Preservation of Closeness," a static construal (Núñez, Edwards, & Matos, 1999; Núñez & Lakoff, 1998). As we shall see, this was reflected in participants' embodied discourse, where a single expression was produced on multiple occasions with gestures that varied in dynamicity.

Gestures co-produced with these targeted lexical items were then classified as dynamic, static, or ambiguous.[2] Because gestures are co-speech motor actions, they necessarily involve dynamic movement. To differentiate between gestures for which dynamism was an artifact of gesture production, and those gestures for which dynamism was truly expressive of mathematical thinking, we devised a coding scheme that attended to details of the motion and timing of the co-speech gesture. A gesture was coded as: (a) dynamic if it used smooth, unbroken motions; (b) static if it consisted of staccato strokes and segmented motions, or of a smooth motion bookended by staccato gestures; and (c) ambiguous if it did not fit into either of these categories.

The graduate students in our study were highly skilled, and thus it was no surprise that every dyad arrived at a reasonable solution that included most of the elements of a complete proof. Half of the dyads even finished early, before the end of the allotted time. In all cases, participants' proofs followed the same outline as the proof supplied by the mathematician who proposed the problem. All six dyads made extensive use of diagrams—primarily graphs of "generic" increasing functions, or sections of functions—in combination with extensive symbolic inscriptions. However, the sequence in which the various modalities were used, including speech, symbolic and graphical inscriptions, gesture and gaze, varied a great deal within and among the pairs of participants. Some participants spent several minutes looking at the paper on which the conjecture was printed or staring into space, others immediately sketched a graph on the blackboard, while several initially discussed what they knew or could immediately conclude about the conjecture (see Edwards & Harel, 2009). These discussions were always accompanied by gestures, some of which were used pragmatically to manage the interaction (turn-taking, emphasis, etc.) while others occurred in conjunction with the use of content-rich mathematical speech.

Examples of Co-speech Representational Gestures

As noted, gestures used to represent the conceptual content of the mathematical discourse were the focus of our analysis. To give a flavor of the range of these representational gestures, below are several examples of dynamic and static representational gestures co-produced with similarly coded lexical items.

Dynamic Gestures

The technical concept of "increase" offers an example of a mathematical construct that seemed to demand an exclusively dynamic treatment in gesture. Participants produced fourteen representational gestures that co-occurred with the lexical items "increasing" or "increase." All of the co-produced gestures were dynamic.

In Figure 10.1, the participant is saying "increasing sequences" and while doing so produces a smooth, unbroken motion, co-timed with his speech. At the onset of the word *increasing*, he begins to fluidly move his left hand upward and toward the right, with his thumb pointing in the direction of motion (Figure 10.1A). As he reaches the end of the word, the motion of his left hand slows slightly while his right hand begins to accelerate, once again moving upward and to the right in the direction of his extended thumb. As he begins to say "sequence," his right hand reaches its top speed (Figure 10.1B). Both hands begin to slow to a stop (Figure 10.1C), and their retraction is co-timed with the end of the word "sequences." Neither his gaze nor his thumbs are directed toward the blackboard or any inscription on it; rather, the deictic aspect of the gesture is used to indicate the direction that the graph of any increasing function would take (up and to the right).

In Figure 10.2, the student is at the blackboard writing a series of inequalities that contradict the assumption that the function is increasing. As he finishes writing, he steps back from the blackboard, drops his hands to his side, and states, "So that contradicts uhhh increasing" (Figure 10.2A). Precisely co-timed with the onset of the word *increasing*, his right hand flies upward and to the right, with his index finger extended in a prototypical pointing hand shape (Figure 10.2B). As he finishes saying "increasing," his right hand slows, pointing to the right of the blackboard for a moment

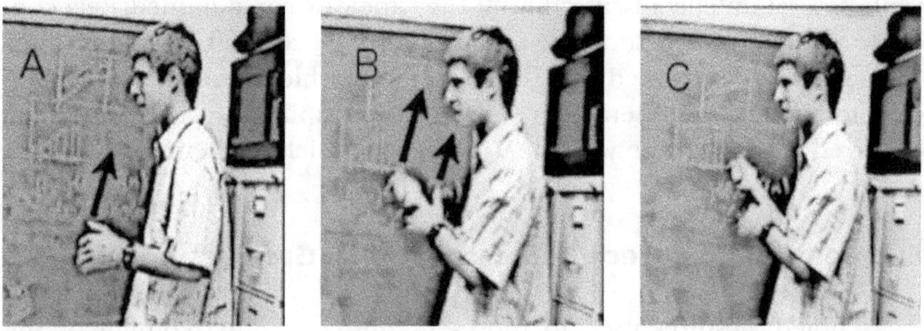

Figure 10.1 Dynamic deictic gesture while saying "increasing sequence." Both hands evoke the sequence's fictive motion.

Figure 10.2 Dynamic gesture while saying "increasing." A fluid rightward hand movement is co-timed with speech.

(Figure 10.2C), before smoothly dropping to his side. This is a quintessential example of dynamic co-speech gesture, consisting of a continuous motion co-timed with a dynamic lexical item.

Static Gestures

Static gestures often accompanied discussions of closeness and containment. In Figure 10.3, the student is discussing the values of a function and wants to consider only those values contained within a restricted region. At first, all her gestures are deictic, anchoring her discussion to the graph

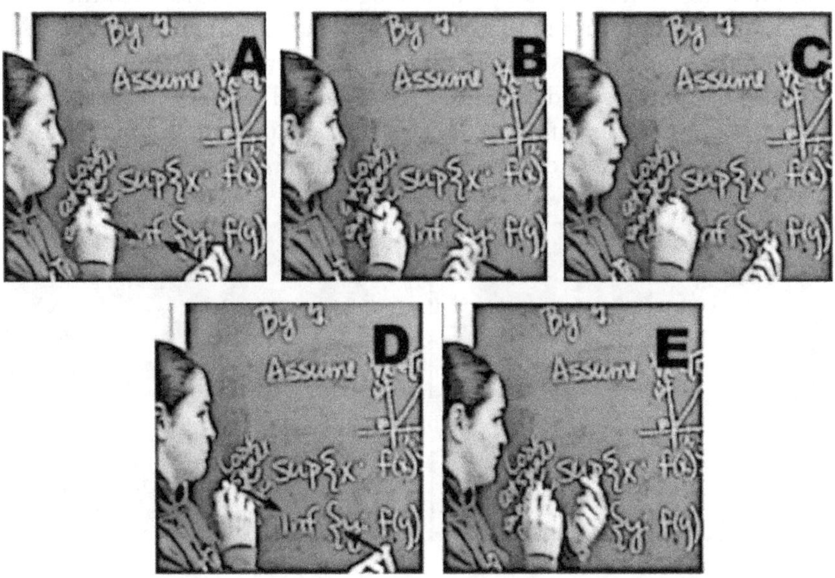

Figure 10.3 A static gesture co-timed with the utterance, "sort of, small enough."

she has drawn on the blackboard. She introduces the region of interest by saying, "Well, if you look at a, sort of, small enough, sort of. . . ." When she begins to describe the region of interest ("Well . . ."), she retracts her hands from the blackboard and positions them in front of her chest, pointing toward each other, and pauses as she says, "sort of" (Figure 10.3A). Co-timed with the onset of the word *small*, she quickly moves her hands toward each other, stopping when they are 10 cm apart, and then quickly retracting her hands to their original distance (Figure 10.3A–C). Co-timed with the word *enough*, she repeats the same inward staccato stroke, stopping abruptly when her hands are at the same distance and retracting once again (Figure 10.3D–E). By indexing two particular points in space with a repeated beat, she evokes the endpoints of a delimited region containing the function's domain. This gesture was thus coded as static because it consists of distinct beats, indexing exact points in space.

Variable Gesture for the Same Utterance

Often a single utterance was amenable to different kinds of representation in gesture. The utterance "to the left," for instance, received both dynamic and static treatments in gesture (Figure 10.4).

In the second frame of Figure 10.4 (labeled B), the participant is producing a static gesture co-timed with the utterance "to the left." One participant suggests to his collaborator that they should look for "implicit continuity to the left of what we're talking about." As he says "to the left," he forms his thumb and forefinger into a U-shape, with the two digits representing the boundaries of a specific interval, performs one forward beat with this hand shape, and then holds his hand still for nearly a full second. This gestural representation of "to the left" captures the static notion of containment within a fixed region to the left of a reference point.

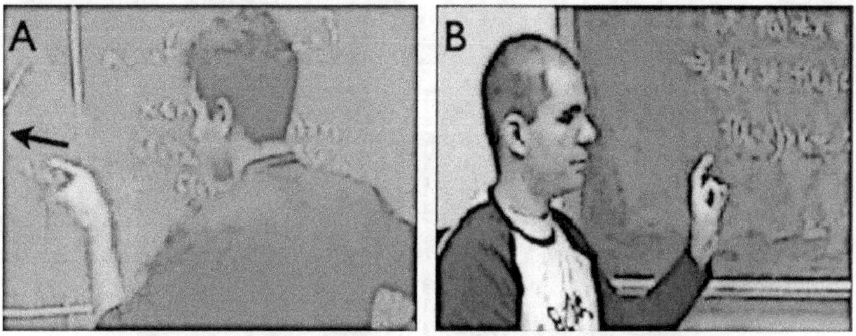

Figure 10.4 Two gestures co-produced with the utterance "to the left." While the gesture on the left is dynamic, the one on the right is static.

The first frame of Figure 10.4 (labeled A) shows the same lexical affiliate ("to the left") co-timed with a dynamic gesture. While discussing the intersection of a function and the line $y = x$, a participant asks whether the function was "going a little bit to the left." As he begins to say "to the left," he points his forefinger to the left and fluidly moves his hand in that direction, retracting his hand as he says "left." In contrast to the example of a fixed (static) region to the left, above, this gesture represents a dynamic sense of the fictive motion of the function as it "moves" in a particular direction.

These two cases illustrate that gestures may share a lexical affiliate but differ in their dynamism. Although highly synchronous with the utterance "to the left," these two gestures exhibit different kinetics and hand shapes.

VARIETIES OF METAPHORICAL GESTURES IN MATHEMATICS

Kendon's (2004) distinctions among types of gestures can help us understand the various ways in which participants used gesture to represent mathematical concepts. As noted above, during enactment the gesture enacts or reproduces some aspect of a physical action. The enactments produced by all of the participants were directly coupled to the environment, for example, re-tracing the path of a graph while holding the chalk barely off the blackboard. Because our focus was on gestures that were not deictic, and these enactments involved coupling gesture with a preexisting blackboard inscription, no enactments were coded in this study. In depiction, the gesturer uses the index finger or some other physical element to outline or "draw" the shape of an object in the air. In the current setting, such "objects" consisted of mathematical entities, given a spatial representation in virtue of their definition, standard graphical inscription, or spatial conceptualization. For example, in Figure 10.3, the participant uses a sequence of staccato beat gestures to delimit a region in space, thus depicting this mathematical "object." Similarly, the second frame of Figure 10.4 shows a U-shaped gesture bounding a region "to the left." It should be noted that these gestural depictions of numerical intervals depend on a preestablished conceptual mapping between number and space, so that a "region" can be interpreted as a bounded portion of the real number line. This conceptual mapping is introduced early within mathematics teaching through representations such as the number line, in which numbers that are "less than" a reference number are conventionally taught as lying to the left of the reference number (Dehaene, Bossini, & Giraux, 1993; Edwards, 2009; Lakoff & Núñez, 2000; Núñez, 2011; Núñez & Marghetis, to appear). Similarly, in the two examples of gestures co-produced with the lexical affiliate increasing, both participants exhibited a marked rightward hand trajectory, mirroring the contingent fact that, historically, graphs of functions have been drawn from left to right. In general, the process of associating number and space can be accounted for by the cognitive mechanisms of conceptual blending and metaphor (Núñez, 2011), and a similar conceptual association between number and space has been shown in the context of arithmetic reasoning (Knops, Thirion, Hubbard, Michel, & Dehaene, 2009; McCrink, Dehaene, & Dehaene-Lambertz, 2007; Núñez & Marghetis, to appear; Pinhas & Fischer, 2008).

In Kendon's category of modeling, the gesturing body part stands in for an object. Both specimens of dynamic gesture shown in Figures 10.1 and 10.2 involve the modeling of a mathematical concept via a fictive trajector. In this case, the trajectory is the increase of the mathematical sequence under discussion, and fictive motion involves the motion of an imagined trajector along this trajectory. In Figure 10.1, the moving hands model the dynamic increase of the sequence, with the hands standing in for the fictively moving trajector. In Figure 10.2, the hand models the trajector that travels along the path of a function, passing through the function's increasing values. Note that the modeling of a trajector moving along the path of a graph is a different representation than enacting the action of drawing the graph in the first place; in modeling, the gesture stands in for a (fictive) moving trajector, whereas in enactment, the gesture reproduces, in part, the action of drawing the graph. For both these examples, extended fingers indicate the direction of the trajectory, constituting a kind of vector representation of the trajector's fictive motion—quite unlike any hand shape that might have been used to hold the chalk while tracing a graph on the blackboard.

Kendon's analysis of precisely how a gesture relates to its referent is thus helpful in characterizing how gestures that co-occur with mathematical speech can represent mathematical ideas in different ways. To examine whether this use of gesture to represent mathematical ideas was truly systematic and reliable, we next looked quantitatively at the co-occurrence of dynamic and static gestures with lexical items associated with dynamic and static concepts.

DYNAMIC CONCEPTS, DYNAMIC GESTURES

The participants produced a large number of gestures, the majority of which were deictic or "pointing" gestures, anchored to inscriptions on the blackboard. Deictic gestures appeared to play a number of roles, including maintaining attention during particularly complex deductions and directing the attention of a collaborator to a salient inscription. While others have studied the significance of deictic gestures for mathematical communication and learning (e.g., Alibali et al., 1999; Goldin-Meadow & Wagner, 2005; Goldin-Meadow et al., 2009), here we focus on participants' representational gestures. Every participant but one produced representational gestures that were co-timed with the dynamic and static lexical items of interest. A total of 166 of these co-timed representational gestures were coded, for a mean of 13.8 coded gestures per participant. Of these, the majority of the gestures were coded as dynamic (50.6%); fewer were coded as static (41.6%).

In line with the hypothesis that gesture can express the metaphorical and/or spatial content of abstract mathematical ideas, gesture dynamicity varied according to the co-occurring concept expressed in mathematical speech (Marghetis & Núñez, 2013). Certain concepts were associated with a prevalence of dynamic or static gestures. For example, as shown in Figure 10.5, gestures co-produced with talk of "increase," "continuity," and "intersection" were more often dynamic; those co-produced with talk of "containment" and "closeness" were more often static.

To test the robustness of this association between the dynamicity of mathematical concept and the dynamicity of gesture, we examined those participants who produced gestures that co-occurred with these concepts and calculated the proportion of gestures that were dynamic. We focused on specific mathematical concepts that, based on previous theoretical analyses, were thought to have a construal that was primarily dynamic or primarily static: the dynamic notion of increase—thought to evoke fictive motion—and the static notions of closeness and containment (Lakoff & Núñez, 2000). A between-subjects analysis of variance found that gestures produced while speaking of "increase" were significantly more often dynamic than static [$F(1,10) = 28.90$, $p = .0003$], while participants produced a significantly higher proportion of static gestures while discussing "containment" [($F(1,12) = 6.75$, $p = .0232$] and "closeness" [$F(1,10) = 76.73$, $p < .0001$].

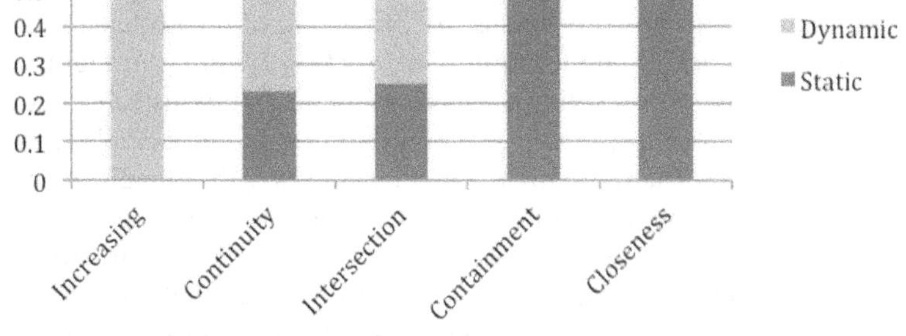

Figure 10.5 Proportion of dynamic and static gestures by co-occurring concept.

Among these expert practitioners, therefore, dynamicity in gesture was reliably associated with mathematical concepts thought to rely on dynamic conceptualization, thus lending credence to the hypothesis that even expert mathematical practice relies on intuitions of motion—a dynamic complement to the static definitions.

DISCUSSION

This empirical study of mathematical practice supports the claim that gesture is an integral component in the production and communication of mathematical meanings during expert proof, and furthermore that gestural meaning-making makes systematic use of dynamicity. Both dynamic and static representational gestures were widespread and recurrent in collaborative mathematical practice among doctoral students. Lexical items linked to mathematical concepts associated with dynamism or fictive motion were systematically paired with dynamic gestures, whereas terms associated with static concepts such as containment were systematically paired with static gestures. In summary, dynamic metaphorical gestures are not restricted to pedagogical or elementary mathematical contexts but are also clearly present in expert mathematical practice.

Because the construction of mathematical diagrams is necessarily a dynamic process, typically involving the tracing of chalk across a blackboard, one might wonder whether the dynamism of gesture in a context involving graphing is a mere echo of the dynamism of the inscriptive motion. Certainly, the motion that inevitably accompanies the creation of a mathematical diagram by hand is probably a factor in the historical and developmental origin of fictive motion for such notions as continuity, function, and limit. The contemporary dynamism of mathematical thought, however, is incredibly robust. Mathematical discourse is rife with fictive motion in the absence of diagrams altogether, as demonstrated by corpus studies of textbooks (Núñez et al., 1999; Núñez & Lakoff, 1998). As an example from the study analyzed here, in Figure 10.2, the participant produced a dynamic gesture not after creating a graphical inscription but after writing a series of static inequalities. The dynamism of his gesture reveals the dynamism of his understanding of increasing sequences. While a graph of a function is visible in Figure 10.1, the participants had not recently attended to the diagram; instead, they were discussing the limit of a function and representing it symbolically using set notation. Moreover, the participant's gesture moves orthogonally to the orientation of the blackboard diagram. The dynamism evinced in gesture is not merely an echo of inscriptive actions, therefore, but evidence for the dynamism inherent in the participants' understanding of the mathematics.

Attention to gesture, and to the body-in-interaction more generally, has opened up new vistas for further research on proof. In the current analysis, environmentally coupled deictic gestures were purposefully disregarded, as when participants pointed at inscriptions on the blackboard. Did this choice ignore an essential component of mathematical gestural behavior? Pointing at the graph of a function, for instance, was not coded, even if the hand dynamically swept along the curve of the graph. This aspect of the coding scheme represented a conscious decision to sacrifice breadth of analysis for experimental traction and focus. The coding scheme successfully discriminated between mathematical gestures that expressed mathematical content (e.g., "The function is increasing") and meta-mathematical gestures that conveyed something about the local practice and environment (e.g., "Let's move the diagram a bit higher on the blackboard"). While deictic gestures are undeniably an important part of mathematical practice, particularly when that practice involves collaboration or explanation at a blackboard (see Goldin-Meadow et al., 2009), the analysis of such environmentally coupled gestures requires care because their interpretation may require attention to the inscription, to the mathematical entity represented by that inscription, or to both (Edwards, 2009; Hutchins & Palen, 1998). By restricting our attention to representational gestures, the analysis avoided this complication. Future research will explore the ways in which environmentally coupled gestures elaborate meaning and direct local practices.

In attending exclusively to formal, disembodied theorems and proofs—products of the sedimentation of local mathematical practices—the study of mathematical thinking has ignored the rich meaning-making practices of flesh-and-blood mathematicians and collapsed multi-agent and multimodal practice into a single idealized agent working within a single written modality. In order to account for the exceptional traits of mathematics—objectivity, necessity, precision, stability—we must remember that, "Of course, in one sense, mathematics is a body of knowledge, but still it is also an *activity*" (Wittgenstein, 2009, § 349).

In this chapter, we have faced off against the received wisdom about mathematics: that it is largely ahistorical, existing independently of human activity; that the real mathematics is the product of mathematical practice—text-based proofs, lemmas, and theorems—while the messy, fleshy activity of actually doing mathematical proof is secondary. Research that takes gesture and other bodily modalities into account is a direct response to this ahistorical tradition and a contribution to an emergent study of mathematics that does justice to actual mathematicians and mathematical activity (e.g., Mancosu, 2008). Without a doubt, expert mathematics is marked by rigorous methods, formal definitions, and symbol manipulation, but mathematical practice—and mathematical proof in particular—requires the mobilization of far more than these abstract resources. Contemporary expert proof, for instance, is a richly embodied practice that involves inscribing and manipulating notations, interacting with those notations through speech and gesture, and using the body to enact the meanings of mathematical ideas. By including gesture in the analysis, we take the mathematician's body seriously as a semiotic resource in the creation and communication of mathematical content. Seen in this light, mathematical proof is no longer an abstract product but is instead a dynamic practice, a human activity that involves talking, inscribing, and, crucially, gesturing: Mathematics is manual labor.

NOTES

1. This research was supported by Spencer Foundation Grant no. 200700151 to Laurie Edwards, Guershon Harel, and Rafael Núñez.
2. Because the goal of the study was to document the deployment of space to represent mathematical content, we excluded interactive gestures—used to manage the communication between the participants.

REFERENCES

Alibali, M. W., Bassok, M., Solomon, K. O., Syc, S., & Goldin-Meadow, S. (1999). Illuminating mental representations through speech and gesture. *Psychological Science, 10,* 327–333.

Alibali, M. W., & diRusso, A. (1999). The function of gesture in learning to count: More than keeping track. *Cognitive Development, 14,* 37–56.

Balacheff, N. (1991). The benefits and limits of social interaction: The case of mathematical proof. In A. J. Bishop, S. Mellin-Olsen, & J. Van Dormalen (Eds.), *Mathematical knowledge: Its growth through teaching* (pp. 175–192). Dordrecht, the Netherlands: Kluwer Academic Publishers.

Bavelas, J. B., Chovil, N., Lawrie, D. A., & Wade, A. (1992). Interactive gestures. *Discourse Processes, 15*(4), 469–489.

Benacerraf, P., & Putnam, H. (Eds.). (1983). *Philosophy of mathematics: Selected readings.* Cambridge, England: Cambridge University Press.

Boero, P. (Ed.). (2007). *Theorems in schools: From history, epistemology and cognition to classroom practice.* Rotterdam, the Netherlands: Sense Publishers.

Cienki, A. (1998). Metaphoric gestures and some of their relations to verbal metaphoric counterparts. In J. P. Koenig (Ed.), *Discourse and cognition: Bridging the gap* (pp. 189–204). Stanford, CA: Center for the Study of Language and Information.

Cornejo, C., Simonetti, F., Ibañez, A., Aldunate, D., Ceric, N., López, V., & Núñez, R. (2009). Gesture and metaphor comprehension: Electrophysiological evidence of cross-modal coordination by audiovisual stimulation. *Brain and Cognition, 70*(1), 42–52.

Dehaene, S., Bossini, S., & Giraux, P. (1993). The mental representation of parity and number magnitude. *Journal of Experimental Psychology: General, 122,* 371– 396.

Dick, A. S., Goldin-Meadow, S., Hasson, U., Skipper, J. I., & Small, S. L. (2009). Co-speech gestures influence neural activity in brain regions associated with processing semantic information. *Human Brain Mapping, 30,* 3509–3526.

Edwards, L. D. (2009). Gestures and conceptual integration in mathematical talk. *Educational Studies in Mathematics, 70*(2), 127–142.

Edwards, L. D. (2010). Doctoral students, embodied discourse and proof. In M. M. F. Pinto & T. F. Kawasaki (Eds.), *Proceedings of the 34th Conference of the International Group for the Psychology of Mathematics Education, Vol. 2* (pp. 329–336). Belo Horizonte, Brazil: PME.

Edwards, L. D., & Harel, G. (2009). Emerging mathematical experts and proof. In M. Tzekaki, M., Kaldrimidou, & C. Sakonidis (Eds.), *Proceedings of the 33rd Conference of the International Group for the Psychology of Mathematics Education* (Vol. 1, p. 370). Thessaloniki, Greece: PME.

Fauconnier, G., & Turner, M. (2002). *The way we think: Conceptual blending and the mind's hidden complexities.* New York, NY: Basic Books.

Font, V., Bolite, J., & Acevedo, J. (2010). Metaphors in mathematics classrooms: Analyzing the dynamic process of teaching and learning of graph functions. *Educational Studies in Mathematics, 75,* 131–152.

Garber, P., & Goldin-Meadow, S. (2002). Gesture offers insight into problem-solving in adults and children. *Cognitive Science, 26,* 817– 831.

Goldin-Meadow, S. (2003). *Hearing gestures: How our hands help us think.* Chicago, IL: University of Chicago Press.

Goldin-Meadow, S., Cook, S., & Mitchell, Z. (2009). Gesturing gives children new ideas about math. *Psychological Science, 20*(3), 267–272.

Goldin-Meadow, S., Kim, S., & Singer, M. (1999). What the teacher's hands tell the student's mind about math. *Journal of Educational Psychology, 91,* 720–730.

Goldin-Meadow, S., Nusbaum, H., Kelly, S. D., & Wagner, S. (2001). Explaining math: Gesturing lightens the load. *Psychological Science, 12,* 516–522.

Goldin-Meadow, S., & Wagner, S. (2005). How our hands help us learn. *TRENDS in Cognitive Sciences, 9,* 234–241.

Goodwin, C. (2000). Action and embodiment within situated human interaction. *Journal of Pragmatics, 32,* 1489–1522.

Hanna, G. (1991). Some pedagogical aspects of proof. *Interchange, 21*(1), 6–13.

Hutchins, E., & Palen, L. (1998). Constructing meaning from space, gesture, and speech. In L. B. Resnick, R. Saljo, C. Pontecorvo, & B. Burge (Eds.), *Discourse, tools, and reasoning: Situated cognition and technologically supported environments* (pp. 22–40). Berlin, Germany: Springer-Verlag.

Kendon, A. (2004). *Gesture: Visible action as utterance.* Cambridge, England: Cambridge University Press.

Kendon, A. (2008). Some reflections on the relationship between "gesture" and "sign." *Gesture, 8*(3), 348–366.

Kita, S. (2000). How representational gestures help speaking. In D. McNeill (Ed.), *Language and gesture* (pp. 162–185). Cambridge, England: Cambridge University Press.

Knops, A., Thirion, B., Hubbard, E. M., Michel, V., & Dehaene, S. (2009). Recruitment of an area involved in eye movements during mental arithmetic. *Science, 324,* 1583–1585.

Knuth, E. (2002). Secondary school mathematics teachers' conceptions of proof. *Journal for Research in Mathematics Education, 33*(5), 379–405.

Lakoff, G., & Johnson, M. (1980). *Metaphors we live by.* Chicago, IL: University of Chicago Press.

Lakoff, G., & Núñez, R. (2000). *Where mathematics comes from: How the embodied mind brings mathematics into being.* New York, NY: Basic Books.

Mancosu, P. (Ed.) (2008). *The philosophy of mathematical practice.* Oxford, England: Oxford University Press.

Marghetis, T., & Núñez, R. (2013). The motion behind the symbols: A vital role for dynamism in the conceptualization of limits and continuity in expert mathematics. *Topics in Cognitive Science, 5,* 299–316.

McCrink, K., Dehaene, S., & Dehaene-Lambertz, G. (2007). Moving along the number line: Operational momentum in nonsymbolic arithmetic. *Perception and Psychophysics, 69,* 1324–1333.

McNeill, D. (1992). *Hand and mind.* Chicago, IL: University of Chicago Press.

McNeill, D. (Ed.). (2000). *Language and gesture.* Cambridge, England: Cambridge University Press

Núñez, R. (2006). Do real numbers really move? Language, thought, and gesture: The embodied cognitive foundations of mathematics. In R. Hersh (Ed.), *18 unconventional essays on the nature of mathematics* (pp. 160–180). New York, NY: Springer.

Núñez, R. (2008). A fresh look at the foundations of mathematics: Gesture and the psychological reality of conceptual metaphor. In A. Cienki & C. Müller (Eds.), *Gesture and metaphor* (pp. 93–114). Amsterdam, the Netherlands: John Benjamins.

Núñez, R. (2011). No innate number line in the human brain. *Journal of Cross-Cultural Psychology, 45*(4), 651–668.

Núñez, R., Edwards, L. D., & Matos, J. F. (1999). Embodied cognition as grounding for situatedness and context in mathematics education. *Educational Studies in Mathematics, 39,* 45–65.

Núñez, R., & Lakoff, G. (1998). What did Weierstrass really define? The cognitive structure of natural and epsilon-delta continuity. *Mathematical Cognition, 4,* 85–101.

Núñez, R., & Marghetis, T. (to appear). Cognitive linguistics and the concept(s) of number. In R. Cohen-Kadosh & K. Dowker (Eds.), *Oxford handbook of numerical cognition.* Oxford, England: Oxford University Press.

Pinhas, M., & Fischer, M. H. (2008). Mental movements without magnitude? A study of spatial biases in symbolic arithmetic. *Cognition, 109,* 408–415.

Rasmussen, C., Stephan, M., & Allen, K. (2004). Classroom mathematical practices and gesturing. *Journal of Mathematical Behavior, 23,* 301–323.

Talmy, L. (2000). *Toward a cognitive semantics: Concept structuring systems* (Vol. 1). Cambridge, MA: MIT Press.

Wittgenstein, L. (2009) *Philosophical investigations* (rev. 4th ed.) (P. M. S. Hacker & J. Schulte, Eds.; G. E. M. Anscombe, P. M. S. Hacker, & J. Schulte, Trans.). Chichester, England: Wiley-Blackwell.

CHAPTER 11

BLENDING ACROSS MODALITIES IN MATHEMATICAL DISCOURSE

Nathaniel J. Smith
University of Edinburgh

When people engage in face-to-face discourse about mathematical topics, they deploy a variety of communicative modalities, including at least speech, gesture, and various forms of symbolic and diagrammatic marking on paper or blackboards. These modalities are not parallel streams: To be parallel means to never meet, but speakers mix these streams constantly, coordinating distinct representational resources in a fine-grained way within a single, shared, and meaning-laden space, and it's in this space that communication and thinking are performed. There are even cases where the apparent boundaries between modalities break down entirely. To understand mathematical discourse, then, it's not enough to examine individual meaningful actions—single utterances, gesture strokes, and so forth. We also need to understand the structure of the space in which they occur and how this structure is built, maintained, and manipulated through communicative acts. In this chapter, I'll examine this question through a series of examples drawn from a corpus of naturalistic video recordings of upper division and grad-

Emerging Perspectives on Gesture and Embodiment in Mathematics, pages 247–264

uate mathematics students working together in small groups to complete their assigned homework in topology and abstract algebra (Smith, 2003).

REAL-SPACE BLENDS

The theory of real-space blending was originally developed for the analysis of sign language (Liddell, 2003) but has proven useful in the study of both gesture (Parrill & Sweetser, 2004) and complex conceptual artifacts like maps and compasses (Hutchins, 2005). This theory builds in turn on the general theoretical framework of conceptual integration theory (Fauconnier, 1997; Fauconnier & Turner, 2002). The core idea of conceptual integration theory is that humans constantly engage in the creation of new conceptual structures, and that they often do this by following a particular recipe: They take two or more existing concepts, identify mappings between their parts, and then blend them together to create a new, indivisible concept whose structure is drawn from the input concepts, adjusted as necessary to produce a coherent whole. Such a process seems to be needed to explain phenomena as diverse as adjective/noun combinations (Sweetser, 1999), *if* statements (Dancygier & Sweetser, 2009), moral judgments (Coulson, 2001), and the development of abstract mathematical concepts (Edwards, 2009; Lakoff & Núñez, 2000). Here, though, we will focus on real-space blends (i.e., those that instantiate a mapping between abstract conceptual structure and the physical structure of the real environment).[1] For example, consider the diagram drawn in Figure 11.1[2]:

(1) the logarithmic spiral...

draws a spiral on the board, Figure 11.1

In physical terms, all the speaker is doing is placing some bits of colored dust on the surface of a whiteboard. But of course the reason he does this is because he is setting up a coherent mapping between this dust's particular spatial arrangement and a certain mathematical object. This object—the *logarithmic spiral*—consists of those complex numbers that can be written as $e^{\theta}\ e^{i\theta}$ for some real number θ. If we conceptualize the set of all complex numbers as forming a two-dimensional plane (itself a complex many-input blend), then the points that satisfy this formula will form the shape of a spiral around the origin. This conceptual spiral is then mapped onto the physical markings on the whiteboard—a second blend, and the one that is the focus of our attention here. Of course a number of accommodations must be made: The whiteboard is not infinite in extent; his mark does not extend infinitely in either direction; his drawing is not accurate in its proportions. Nonetheless, a great deal of internal structure does map between these domains:

Figure 11.1 "The logarithmic spiral..."

The various locations on the drawn spiral correspond to different points on the abstract spiral, the blank space around the drawn line maps to other complex numbers, and the center of the spiral maps to the plane's origin. The speech provides a critical cue for what the marking is intended to mean and together with the shape of the drawn spiral allows us to construct the appropriate mapping. But once the blend is established, it remains indefinitely, lingering long after the speech is gone. In fact, immediately after this spiral is drawn, the group turned to discuss a tangential topic for several minutes before mentioning or interacting with this drawing again.

We'll follow their example and return to the spiral below after examining a gesture example.

THE LIFE CYCLE OF REAL-SPACE BLENDS

In (2), the speaker on the right describes how she constructed a new set (let's call it S_2) by performing a series of operations on some original set (S_1, say), with reference to a third set she dubs U. Overall, she has two points she wants to make: The first is that S_2 has a simple relationship to S_1; it is obtained by taking all elements of S_1 that are not in U or in set arithmetic terms, $S_2 = S_1 - U$. The second is that the new set S_2 has a certain topological property. In topology, sets can be open, closed, both, or neither, and these properties are defined in part by how they propagate through the two operations of set complement and set union. Here, she demonstrates that if S_1 is closed and U is open, then S_2 is also closed. To accomplish this, she describes a complex series of complement and union operations, which together show that S_2 is closed. Simultaneously, she uses gesture to build a complex real-space blend to track these operations and show that this sequence of operations is equivalent to a simple set difference operation.

Because they are doing topology, the sets are conventionally conceptualized as subregions of some larger region of space. She begins with the set S_1, blended onto her right hand (Figure 11.2a). Notice that this is essentially similar to the process described above, with the physical structure of her hand replacing the whiteboard marker pigment[4]:

(2) so [$_1$**my** original closed set,
puts right hand up in air, holds, Figure 11.2a
right, was [the comp**lement of some** open set.]
scoops left hand around to indicate open set, Figure 11.2b
now I'm [taking **that** open set,]
repeats left hand scoop, quicker this time, Figure 11.2c
union [another open set... right?]
left hand hold, Figure 11.2d
[**that open set**]
left hand scoop again, Figure 11. 2e
union [$_2$**U**,
left hand hold again, Figure 11.2f
and taking the]$_1$ [**complement of that**
right hand moves away from left, Figure 11. 2g
]]$_2$...
both hands drop

First we see that by initially placing the right hand, she has created a blend that extends beyond it—she is also structuring the space around her hand so that left-hand gestures she performs in the physical region where her right hand is not located refer conceptually to the points that are not in the set S_1 (i.e., the set complement of S_1) (Figure 11.2b, c).

She then introduces the set U, which she conceptualizes as partially overlapping the original set S_1.[4] This set is also blended onto the left hand (Figure 11.2d), and the set overlap is represented by placing the corresponding gestural hold physically into the region between her right-hand index finger and thumb so that the convex hulls of the two hands representing S_1 and U physically overlap.

In Figure 11.2e and f, she alternates between invoking the complement set and the new set U; because she has only two hands, she cannot perform

(a) so [my original closed set

(b) right, was [the complement of some open set]

(c) now I'm [taking that open set,]

(d) union [another open set... right?]

(e) [that open set]

(f) union U

(g) and taking the] [complement of that]

Figure 11.2 "So my original closed set."

them simultaneously while also maintaining the right-hand hold as a reference point anchoring the blend. The speech co-timed with both of these gestural pairs c–d and e–f tells us what the relation between these sets is: She is taking the union between them (i.e., the set that contains all points that are in either or both of the original sets). Interestingly, her syntax here does not follow the conventional rules of English grammar; while it is embedded in a standard English sentence, the phrasing *X union Y* instead follows the conventions for reciting mathematical formulas like $X \cup Y$.

According to these conventions, we read symbols from left to right, and this rule is what ultimately provides the temporal ordering on her gestures—the order of her gestures, first the complement, second the new set *U*, makes clear that the two gestures correspond to the two parts of such a formula.

Once this is established, yet another new set enters the discourse: the set that is the union of the complement and *U* and that we know by the definition of a union should, in the blended space, consist of the region that overlaps precisely with the sets used to construct it. Starting in Figure 11.2g, then, the left-hand hold becomes ambiguous: It has been established that it occupies the same space as *U*, and *U* occupies the same space as the union set; therefore, the left-hand hold also gives us the blended position of (part of) the union set.

Finally, she takes the complement of this union set. For depicting this final set, she uses the right hand, finally releasing it from the position where it has been held since the beginning of this description—it moves slightly from the S_1 position to a new position that is almost identical except that it no longer overlaps with the left hand representing *U*. This motion encodes what the speech does not: that this set is exactly the set we originally described, $S_2 = S_1 - U$, and that this sequence of operations together has performed a set difference.

This motion would not be meaningful without real-space blend that it occurs within, and this blend required the whole sequence of preceding speech and gesture to construct and elaborate. But once the rich blended space has been established, the simple motion in Figure 11.2g is enough to convey a proof of an interesting theorem: that a closed set minus an open set is a closed set. The purpose of the preceding speech and gesture is, in some sense, to put this blend in place and allow this final gesture.

If we look at this sequence of speech and gesture, we see that it constantly interacts with the blended space in complex ways. First, the blended space is first established in Figure 11.2a. Then the left-hand gestures and the final right-hand gesture all depend on the existing blend to depict new content (e.g., the complement set gestures, which are interpretable only by comparing their spatial position to the spatial position of S_1) while elaborating it further (e.g., by placing *U* so that it partially overlaps S_1). Simultaneously, while this is occurring, she devotes some energy to maintaining the blend. This is accomplished through an elegant dance of hold gestures: Whichever hand is not currently elaborating the blend is held steady, anchoring whichever portion of the blend it is currently assigned—first she uses the right hand, then both (briefly) in Figure 11.2f, and finally the left hand takes over for Figure 11.2g (Enfield, 2003, 2009). These holds both signal the continued relevance of the blend and provide a visible spatial reference point for it. Finally, once the blend has served its purpose, she drops both hands, signaling the end of this blend's utility, and freeing up the physical space to whatever content is needed next.

These operations—construction, maintenance, exploitation/elaboration, and destruction—form the fundamental life cycle of real-space blends in discourse, which interlocutors must systematically manage to accomplish their goals. Different strategies may be available and used depending on the specific details of how each blend is instantiated, the current goals, and trade-offs between conflicting ones. For example, a whiteboard drawing remains until erased, but gestures are more transient, and thus ambiguity is more likely to arise about whether an earlier blend is still relevant; the speaker in (2) finds one solution to this problem using hold gestures, but this solution creates a new problem of how to simultaneously maintain and elaborate the blend given her limited number of hands, which she in turn solves with the switch in Figure 11.2g. The point of this schema is not to allow us to predict exactly what interlocutors will do in any particular situation, but to provide a conceptual framework to understand these competing constraints on the structure of the persistent, meaningful spaces that discourse participants navigate to succeed—and to encourage us to consider all the work going on outside of the individual gestural strokes.

REAL-SPACE BLENDS AS SHARED SPACES

Real-space blends, by definition, map physical structure and relationships onto conceptual structure and relationships and vice versa. The previous examples demonstrated real-space blends in two different modalities separately—drawing and gesture—together with a third modality, speech. But crucially, real-space blends are not modality-specific. While these modalities each have their own unique conventions and affordances (Edwards & Robutti, this volume), their physical components all take place within the same physical environment and a shared spatial and temporal frame. A real-space blend lives in this shared physical space and is thus accessible to all modalities simultaneously. At this level of analysis, hand positions, drawn lines, and written text are simply different ways of locating meaning in space; and hand motions, acts of drawing and writing, and speech utterances are ways of locating meaning in time (Hiraga, 2005; Sweetser, 2006). Spatial and temporal relationships can hold between any and all of them and are used to link together these disparate representational tools into a coherent, meaningful blended space.

Familiar examples of this phenomenon include the co-timed production of speech and gesture—a temporal relationship indicating a shared meaning—or a pointing gesture directed at a diagram—a spatial relationship used to indicate some conceptual entity. In (3) and (4), our interlocutors return to the diagram drawn in (1) and produce an elaborate complex of representations around it involving speech, gesture, writing, and drawing.

The problem they are working on requires that they find a function that sends all the points on the complex plane onto the logarithmic spiral, and that also satisfies some other conditions.[5] One participant begins by proposing a candidate function:

(3) I'm thinking we're gonna send all the [**circles**]
both hands up and rounded to form circle, Figure 11.3b
. . . like e to the t e to the i theta are all going to go to these
writes "$e^t e^{i\theta}$" *writes "→" pointing at "$e^\theta e^{i\theta}$," already on the board, Figure 11.3c*
you know what I'm sayin?
moves to other side of board
like . . . all the circles[**. . .**]
traces a circle on board over spiral with pen, but doesn't draw, Figure 11.3d
are gonna go to[**. . .**] the **point** on the circle that
points at "$e^\theta e^{i\theta}$," Figure 11.3e
points in the air with the tip of the pen
is [**e to the theta times e to the i theta**]
jabs the air four times, approximate 2-d positions shown in Figure 11.3f In 3-d, it's clear the positions are laid out $1^2 3^4$*, schematically similar to* $e^\theta e^{i\theta}$.

His proposed function maps all the points in each circle centered on the origin to the point where that circle intersects the logarithmic spiral. To explain this idea, he produces a flurry of different representations. I count six representations of the circles:

- the phrase *all the circles*;
- an iconic circle gesture (Figure 11.3b);
- the phrase *e to the theta e to the i theta*, co-timed with:
- writing $e^t e^{i\theta}$, the formula for a circle with radius t;
- the phrase *all the circles* again;
- tracing a circle in the air over the spiral (Figure 11.3d, again taking advantage of a visible entity to anchor a larger blend; here the presence of the spiral implies the complex plane around it).

Three representations of the mapping:

- the phrase *are all going to go to*,
- co-timed with an arrow drawn on the board;
- the same phrase again but reduced, *are gonna go to.*

(a) *all the* ***circles***

(b) *e to the t e to the i theta are all going to go to these*

(c) *all the circles. . .*

(d) *go to[. . .]the point*

(e) ***e to the theta times e to the i theta***

Figure 11.3 "We're gonna send all the circles."

And seven representations of the intersection point:

- the anaphoric phrase *these* co-timed with:
- the drawing of an arrow pointing at the text $e^{\theta}\ e^{i\theta}$;
- a gestural point at this text (Figure 11.3e);
- an iconic gesture co-timed with:

- the word *point*;
- the phrase *e to the theta times e to the i theta* co-timed with:
- a schematic representation of the spatial layout of this formula gestured into the air in front of him (Figure 11.3f).

Several of these actions depend on preexisting real-space blend structure in his environment: the writing of the formula and its arrow point to an existing formula (Figure 11.3b), the gestural point to this same formula (Figure 11.3d) and the gesture over the existing spiral diagram (Figure 11.3c). Notice that these are performed without any regard for the apparent boundaries between modalities. It's possible that for some purposes we might want to analyze the lower portion of the whiteboard, with its equations, as a different space from the upper portion, with its diagrammed spiral, and different again from the iconic circle gesture performed in peripersonal space (Figure 11.3a). But such a division is entirely orthogonal to the division between modalities: In all of these spaces, we see the same principles of spatial and temporal relationships being used to bind meaning across modalities that include speech, gesture, writing, and drawing (Goodwin, 2007; Haviland, 2000, 2007).

This explains some key parts of how the many representations we counted above gain their individual meanings. But how is it that he produced sixteen representations and yet communicates a single coherent idea of circles mapping to points according to a single rule? How does he bind these different representations together (Nathan & Alibali, 2011)? Some of the linking is done locally by co-timing and by the word *like*, which indicates syntactically that the meaning of the phrase *e to the theta e to the i theta* is an expansion/clarification of the meaning of the phrase *all the circles*. But then overlaid on this local structure is a larger repetition (Du Bois, 2010). The phrase *you know what I'm sayin?* and his movement from the left to the right half of the frame divide the explanation into two parts. Both parts first describe the circles, then their mapping, and then the points they map to. This structural repetition, together with individual repeated fragments (*all the circles* versus *all the circles, are all going to go to* versus *are gonna go to*, the repeated use of the written e^{θ} $e^{i\theta}$), make clear that the two overt descriptions in fact refer to the same underlying concept and similarly for the corresponding pieces of the two descriptions. Note that these are repetitions of physical form—when he repeats a phrase, we notice the repetition because he literally makes similar sounds with his mouth. That is, this is yet another relationship between forms that is mapped into relationships between meanings via real-space blending. By tracing multiple such formal relationships, we come to understand that the circle gestures in Figure 11.3a and c refer to the same entity, and the formula in Figure 11.3b likewise.

Another interesting possible example of form repetition used to link concepts occurs purely within the mathematical notation used in this example. The scope and quantification of the variable θ is quite ambiguous. For $e^t e^{i\theta}$ to be a circle, t must be fixed while θ varies; for $e^\theta e^{i\theta}$ to be a point, θ must be fixed. As a result, there's no mathematical formalism that lets us define θ so that the formula $e^t e^{i\theta} \to e^\theta e^{i\theta}$ makes sense as a mapping of a circle to a point. Yet this doesn't seem to bother the participants. Perhaps they note that the formulas are identical except for the first exponent and understand this as defining an identity constraint, just like the speech's syntactic repetition creates identity mappings between analogous elements: For each circle with its varying θ, pick the value of θ where $t = \theta$.

Returning to the episode illustrated in Figure 11.3, thirty seconds later, the original speaker and one of his collaborators begin to find reasons that the mapping he has suggested cannot fulfill the requirements set out in their homework problem:

(4) **R** the pre-image of the [**origin can't just be** the] origin
points with pen at center of spiral on board, multiple beats

. . .

so that can't be right, like
we can't– [we can't be sending all the circles] to–
draws a circle on the board, along same path as had gestured a circle earlier, Figure 11.4a

L cause it seems like you're like [nailing down **all this stuff here**
O hand taps down along spiral, Figure 11.4b
and then you're gonna have to **break something apart too**] <XXX>
hand stops moving, uses fingers as pincers to "pull" bits of space toward the spiral, Figure 11.4c

R well I was thinking like we could send [all of the **circles**]
index finger traces previously drawn circle
[to the unique **point** that this thing intersects]
index finger stops on intersection of drawn circle and drawn spiral, Figure 11.4d
cuz [the **spiral**] intersects a circle at one and only one point
traces spiral in air in front of him with index finger

Let's compare Figure 11.3c to Figure 11.4a. Both occur co-timed with the initial portion of schematically similar utterances, *all the circles... to the point.* (In the second case, he is interrupted halfway through this utterance, but we can see what he intended by how he resumes the utterance at the end of the example.) Both involve similar motions: tracing the pen in a circular motion over the spiral diagram. The difference is that in the first case, the pen is held just above the surface of the whiteboard so that no mark is made; in the second case, the pen is lowered slightly so that a trace is left behind, and the motion is somewhat slower and more deliberate. Why is this?

Consider that for the students' purposes, the precise size of the circle is irrelevant—it's a generic representative of the class of all circles that share the same center, so the motions could just as easily have been a bit bigger or smaller. A gesture, being both schematic and transient, leaves vague the exact size of the circle that is intended. So while the gesture establishes that there is some point where the circle and spiral intersect, it does not establish which precise physical, spatial location that point should be blended onto, and in particular, leaves no way to later determine whether any particular location corresponds to this special point. The first time this is irrelevant: While the student immediately produces a reference to this point, his elaboration uses the formula (pointing at $e^{\theta}\ e^{i\theta}$, saying *e to the theta times e to the i theta*), not the diagram. The second time, though, when he eventually gets a chance to finish his sentence and say *to the unique point*, he now elaborates by directing a gesture at the particular location in the diagram where the circle and spiral lines intersect. This reference succeeds only because the circle he has drawn has elaborated the spatial structure of

(a) *<u>all the circles</u>*

(b) *nailing down* ***all this stuff here***

(c) ***break something apart too***

(d) *the unique* ***point***

Figure 11.4 "We can't be sending <u>all the circles.</u>"

the blend, creating a stable and unambiguous perceptual anchor for this intersection point, in a way that the gesture did not.

Clearly his reasoning did not proceed as follows: first, deciding which modality to use (gesture or drawing); and second, deciding what kind of gesture or drawing to make. In fact, the drawing has more canonical "gesture-like" features than the actual gesture, including both a pre-stroke hold and simultaneous production with a co-referential syntactic constituent. Instead, it appears as if he decided that he wanted to exploit this real-space blend to spatially describe an abstract circle and then compared the affordances of these two different methods of tracing the circle: pen up is quicker, more vague, and more transient, thus cluttering the diagram less; pen down is slower, but has the side-effect of leaving a precise and stable anchor for the elaborated structure. Which choice is optimal in any particular situation depends not on the *current* referential goal—they both refer to the same circle via the same real-space blend—but on whether having a stable anchor will be useful for future actions. In each case, he does in fact choose the option that matches the future actions he does in fact perform. It seems unlikely that this is an accident. Rather, I would argue, it demonstrates the strategic selection between different strategies for manipulating the blended space on the basis of current discourse goals and plans for achieving those goals.

Finally, consider the interruption. For our purposes, the interesting thing about this is that speaker L simply moves into the real-space blend spaces and uses speaker R's diagram as a substrate for making her own statements, in a way reminiscent of what Furuyama (2000) refers to as "collaborative gestures," in which one person reaches into another's gesture space to manipulate the structure of their blend directly. These blends are shared not just between modalities but also between interlocutors.

In (5) we see something that blurs the line among gesture, writing, and drawing to invisibility—what one might call "gestural drawing." Here the speaker starts with markings on her paper corresponding to a group T

Figure 11.5 **"pshew pshew pshew."**

that is "acting" on a point. The idea of a group acting on a point is that each element of the group somehow modifies the point, so that from one point arises a whole collection of differently modified points.

(5) **L** you're taking it [**up...**]
draws an arrow showing the movement of a point
R you're moving it somewhere else
L acting on it with *T*, [**...pshew pshew pshew pshew pshew pshew**]
repeated strokes across paper from T to point, Figure 11.5
and then bringing it back.

Speaker **L** draws multiple lines from the place corresponding to the group toward the place corresponding to the point. These lines are drawn in a fast and haphazard fashion, which makes no sense in terms of drawing because a line is the same whether drawn quick or slow—but if one conceptualizes the process as a series of group elements hurtling out of *T* and smacking into the point, then the motion does code appropriate force dynamics and imagistic structure. Furthermore, they are drawn on top of or nearly on top of each other, which again makes little sense if one's motivation is simply to draw a particular picture, but it does accurately represent the motion of multiple elements whose paths share start and end points. It's even unclear whether this is drawing or writing because arrows are such a common conventionalized mathematical notation, and notational arrows can do all the normal things arrows can do in diagrams. Is this gesture or goal-directed action? In this instance, the distinction doesn't seem useful.

Meanwhile, she uses speech to produce co-timed sound effects, which iconically represent repeated fast metaphoric motion. Here the overall timing is driven by an imagined activity, and both the hand movements and speech are aligned to and represent it. The standard terminology for real-space blends seems particularly inadequate here because "space" is so clearly just one aspect of the physical environment that is participating in the blend (a residue of the theory's sign-language heritage). Perhaps it's a real-space-and-sound blend.

THINKING IN A REAL-SPACE BLEND

Interlocutors share a physical environment but, lacking telepathy, do not directly share a mental environment. This makes real-space blends, which map rich mental constructs into the shared physical environment, useful for accomplishing communicative goals. Theoretically, though, this is not a requirement; real-space blends can and do exist outside of communicative contexts. In some ways, this observation is commonplace: Of course people draw diagrams when solving problems alone, and many cultural artifacts' entire purpose is to encode meaningful structure into their physical structure: clocks, maps, slide rules, protractors, calculators, and so forth (Hutchins, 2005).

Because real-time interactional communication uses the same cognitive tool of real-space blending, it is perhaps not a surprise that people also use ad hoc, iconic/metaphoric gestures to support their own thinking (Chu & Kita, 2008; Rodríguez & Palacios, 2007). In (6), the two students are working independently in their notebooks. Then one of them pauses her writing, raises her hands in a hold gesture, and remains motionless for thirty seconds before resuming writing.

(6) **R** ... [. . .] ...
30 second silent hold, Figure 11.6

At no point does her partner look up or interact in any way. This appears to be an anchoring gesture, like those we saw in (2), used to maintain a

Figure 11.6 "...[. . .]..."

Figure 11.7 "*It's [basically it's like taking an* ***open interval.***"

blend that she is using for some cognitive purpose of her own. While the absence of speech makes it difficult for us as analysts to know what exactly this gesture means to her, it's intriguing to observe that this gesture is similar to one that she produced approximately one minute earlier while they were still discussing the problem.

(7) **R** yeah it's [basically it's like taking an **open interval**
forms a two-handed hold, Figure 11.7
rotates both hands back-and-forth, making a beat

While this occurred a minute earlier, it is the last-but-one utterance before they turned to their notebooks to work independently and highly suggestive that the real-space blend she uses for cognitive purposes in (6) is actually the same real-space blend she used earlier for communicative purposes in (7). If the same blend can be used for both purposes, then we have a new analytic puzzle: When we observe people using such representations within a problem-solving discourse, how can we tell whether they are being used for communicative (other-directed) or cognitive (self-directed) purposes? Many meaningful actions might even serve both purposes simultaneously.

CONCLUSION

A real-space blend is a structure-preserving map between physical and conceptual realities, and it is a generic cognitive tool that underlies many different modalities of meaningful action, including at least speech, gesture, writing, and drawing. These blends often persist through time and allow for rich, fine-grained interaction between modalities and between interlocutors, and they support both communicative and cognitive activity. To be useful, though, they must be managed: Blends must be constructed, maintained, elaborated, and deconstructed, and many discourse actions are systematically chosen to accomplish these tasks simultaneously with accomplishing more immediate communicative goals.

When looking at a single gesture in isolation, the stroke catches our attention. But in the context of an ongoing discourse, every word, gesture, and stroke of the pen is performed into a physical environment whose structure has been richly overlaid with meaning, and every such action also further modifies this environment, affecting future actions. In many cases, it may be more important to understand the space in which a gestural stroke occurs, and the way in which it modifies that space, than it is to understand the stroke itself.

NOTES

1. Because co-present interlocutors have perceptual access to their shared physical environment, we can treat their cognitive representation of this environment as an especially detailed and concrete kind of concept and use it as an input to the blending process. See Liddell (2003) and Hutchins (2005) for further discussion of this point.
2. For transcripts, I use ellipses to represent silent pauses and "<XXX>" to denote inaudible speech. When two speakers interact, they are labeled "L" and "R," corresponding to their position on either the left or right of the video frame according to the reader's viewpoint. Gestures are represented by [brackets] to indicate the timing of pre- and post-stroke holds, bolding to mark the gestural stroke itself, and italicized text the motion that is horizontally aligned to the start of the stroke. Underlining marks periods in which the speaker is writing or drawing simultaneously with his or her speech.
3. Subscripts on brackets are used to disambiguate brackets that mark the start and end of two partially overlapping hold gestures.
4. Interestingly, the overlap occurs only in the gesture not in the speech. This corresponds to the fact that formally the degree of overlap between S_1 and U is irrelevant: The proof goes through just as well even if S_1 is a subset of U or if they don't intersect at all. But in either of these cases, the result is trivial (because either S_2 is the empty set, or else $S_2 = S_1$). Her speech describes the general case, just as a formal written presentation would. Her gesture not only helps by tracking the entities involved but also by depicting the particular case where the proof is most interesting and requires real thought.
5. Specifically, it must be a "retraction"; they eventually prove that no such function exists.

REFERENCES

Chu, M., & Kita, S. (2008). Spontaneous gestures during mental rotation tasks: Insights into the microdevelopment of the motor strategy. *Journal of Experimental Psychology: General, 137*(4), 706–723.

Coulson, S. (2001). *Semantic leaps: Frame-shifting and conceptual blending in meaning construction.* Cambridge, England: Cambridge University Press.

Dancygier, B., & Sweetser, E. (2009). *Mental spaces in grammar: Conditional constructions.* Cambridge, England: Cambridge University Press.

Du Bois, J. W. (2010). *Towards a dialogic syntax.* Unpublished manuscript, UC Santa Barbara.

Edwards, L. D. (2009). Gestures and conceptual integration in mathematical talk. *Educational Studies in Mathematics, 70*(2), 127–141.

Enfield, N. J. (2003). Producing and editing diagrams using co-speech gesture: Spatializing nonspatial relations in explanations of kinship in Laos. *Journal of Linguistic Anthropology, 13*(1), 7–50.

Enfield, N. J. (2009). *The anatomy of meaning: Speech, gesture, and composite utterances.* Cambridge, England: Cambridge University Press.

Fauconnier, G. (1997). *Mappings in thought and language.* Cambridge, England: Cambridge University Press.

Fauconnier, G., & Turner, M. (2002). *The way we think: Conceptual blending and the mind's hidden complexities.* New York, NY: Basic Books.

Furuyama, N. (2000). Gestural interaction between the instructor and the learner in origami instruction. In D. McNeill (Ed.), *Language and gesture* (pp. 99–117). Cambridge, England: Cambridge University Press.

Goodwin, C. (2007). Environmentally coupled gestures. In S. D. Duncan, J. Casssell, & E. T. Levy (Eds.), *Gesture and the dynamic dimension of language* (pp. 195–212). Amsterdam, the Netherlands: John Benjamins Publishing Company.

Haviland, J. B. (2000). Pointing, gesture spaces, and mental maps. In D. McNeill (Ed.), *Language and gesture* (pp. 13–46). Cambridge, England: Cambridge University Press.

Haviland, J. B. (2007). Master speakers, master gesturers: A string quartet master class. In S. D. Duncan, J. Casssell, & E. T. Levy (Eds.), *Gesture and the dynamic dimension of language* (pp. 147–172). Amsterdam, the Netherlands: John Benjamins Publishing Company.

Hiraga, M. K. (2005). *Metaphor and iconicity: A cognitive approach to analysing texts.* Basingstoke, England: Palgrave Macmillan.

Hutchins, E. (2005). Material anchors for conceptual blends. *Journal of Pragmatics, 37*(10), 1555–1577.

Lakoff, G., & Núñez, R. (2000). *Where mathematics comes from: How the embodied mind brings mathematics into being.* New York, NY: Basic Books.

Liddell, S. K. (2003). *Grammar, gesture, and meaning in American Sign Language.* Cambridge, England: Cambridge University Press.

Nathan, M. J., & Alibali, M. W. (2011). How gesture use enables intersubjectivity in the classroom. In G. Stam & M. Ishino (Eds.), *Integrating gestures: The interdisciplinary nature of gesture* (pp. 257–266). Amsterdam, the Netherlands: John Benjamins Publishing Company.

Parrill, F., & Sweetser, E. (2004). What we mean by meaning: Conceptual integration in gesture analysis and transcription. *Gesture, 4*, 197–219.

Rodríguez, C., & Palacios, P. (2007). Do private gestures have a self-regulatory function?: A case study. *Infant Behavior and Development, 30*(2), 180–194.

Smith, N. J. (2003). *Gesture and beyond.* Undergraduate thesis, Program in Cognitive Science, University of California at Berkeley, Berkeley, CA.

Sweetser, E. (1999). Compositionality and blending: Semantic composition in a cognitively realistic framework. In G. Redeker & T. Janssen (Eds.), *Cognitive linguistics: Foundations, scope, and methodology* (pp. 129–162). Berlin, Germany: Mouton de Gruyter.

Sweetser, E. (2006). Whose rhyme is whose reason? Sound and sense in Cyrano de Bergerac. *Language and Literature, 15*(1), 29–54.

ABOUT THE CONTRIBUTORS

Martha W. Alibali is Professor of Psychology and Educational Psychology at the University of Wisconsin–Madison. Her research investigates processes of knowledge change in cognitive development and mathematics learning. A central focus of her work is on the role of gestures in learning and communication, particularly in educational settings. Her prior work has shown that students' gestures reveal important information about their mathematical reasoning processes. Current projects explore how teachers use gestures in classroom settings, how teachers' gestures influence students' learning, and how perception, action, and gesture are involved in mathematical thinking and learning.

Ferdinando Arzarello is Full Professor of Elementary Mathematics from an Advanced Standpoint in the Dipartimento di Matematica "G. Peano", Università di Torino, Italy. His research interests concern Mathematics Education and specifically the following topics: Pre-algebra and algebra learning, learning geometry and calculus in technological environments, embodiment and gestures in mathematics, curricular design, and theoretical frameworks for mathematical learning and teaching. He collaborates with the Italian Ministry of Education for curriculum development and for teacher education programs. He has been President of the European Society for Research in Mathematics Education (ERME) and currently is President of the International Commission for Mathematics Instruction (ICMI).

R. Breckinridge Church is Professor of Psychology at Northeastern Illinois University. She received her doctorate in Educational Psychology

Emerging Perspectives on Gesture and Embodiment in Mathematics, pages 265–271

from The University of Chicago and did post-doctoral clinical research as part of a Michael Reese Hospital/University of Chicago program. She is a developmental and cognitive psychologist who pioneered the study of gesture-speech mismatches as an index of transitional knowledge in children. Her recent work focuses on listeners' abilities to detect information expressed uniquely in gesture, gesture as an instructional tool for learning and gesture's role in the processing of multiple languages. Her research has been funded by the Institution of Educational Science and National Science Foundation grants.

Tommy Dreyfus is Professor Emeritus of Mathematics Education at the Department of Mathematics, Science and Technology Education of Tel Aviv University. His main research interests lie in students' processes of constructing abstract mathematical knowledge, and the influence of context such as the mathematical, curricular, social and technological learning environment on these processes. Together with students and colleagues, he has investigated knowledge constructing processes on topics ranging from fractions via algebra, probability, functions, infinite sets, exponential growth, and integration to dynamical systems; notably, definition and proof have played a prominent role in his research. Dreyfus has used gestures in his research as one of the means for the researcher to interpret students' thinking and hence provide evidence of their knowledge constructing processes.

Laurie D. Edwards is Professor of Education in the Kalmanovitz School of Education at Saint Mary's College of California. Her research interests lie in understanding mathematical thinking, teaching and learning, focusing in particular on the role of the body in conceptualizing and communicating mathematics. She has investigated the use of gesture in conjunction with other communicative modalities, and has brought the tools of cognitive linguistics to bear on understanding mathematical situations. She has also studied how students learn with technology, specifically with exploratory computer microworlds, and is interested in the conceptual impacts of different kinds of mediated mathematical experience. She has served as Secretary and Treasurer of the International Group for the Psychology of Mathematics Education.

Solange Hassan Ahmad Ali Fernandes is lecturer in the Post-Graduate Programme in Mathematics Education of the Anhanguera University of São Paulo, Brazil. Her research focuses on understanding the relationships between the body and mathematical cognition in the cases of learners whose sensory or cognitive characteristics differ from what is considered the norm. Her studies hence attend to the material, technological and semiotic tools which mediate cognition and how these resources transform the processes of teaching and learning those labelled as different.

Francesca Ferrara is currently Associate Professor of Mathematics Education in the Department of Mathematics at Università di Torino, Italy. She collaborated on the Italian translation of the book Lakoff, G. & Nunez, R.E. (2000). *Where Mathematics Comes From: How The Embodied Mind Brings Mathematics Into Being*. New York, NY: Basic Books. Her research interests lie in the role and uses of technology in mathematics teaching and learning; how theories of embodiment have informed and changed our ways of doing research in the field; the role of visualization and imagination in mathematical thinking; and mathematical inventiveness. More recently, she has started to consider how new materialist perspectives offer reflections on the ontology and the pedagogy of the concept.

Susan Goldin-Meadow is the Beardsley Ruml Distinguished Service Professor in the Departments of Psychology and Comparative Human Development, and the Committee on Education at the University of Chicago. Her research focuses on the home-made gestures profoundly deaf children create when not exposed to sign language, and the gestures hearing speakers around the globe spontaneously produce when they talk, with an eye toward what gesture can tell us about how we think. She has discovered that the gestures children produce when explaining their solutions to math problems not only reflect their understanding of the problems, but can also play a role in changing that understanding.

Lulu Healy is a lecturer in the Post-Graduate Programme in Mathematics Education of the Anhanguera University of São Paulo. Her research interests focus on the design of innovative ways of doing and expressing school mathematics and the use of digital technologies in the teaching and learning of mathematics. She is particularly interested in the challenges associated with the building of a more inclusive school mathematics and in understanding the mathematical practices of learners with disabilities. She is currently coordinator of the research group Technology and Means for Expressing Mathematics and director of the research programme, Towards an Inclusive School Mathematics, which aims to investigate the relationships between sensory experiences and mathematical cognition and to design and evaluate mathematical learning scenarios for students with a diverse range of educational needs.

Autumn B. Hostetter is Associate Professor of Psychology at Kalamazoo College in Michigan where she teaches courses in cognition, psycholinguistics, and research methods. In her research, she explores the cognitive underpinnings of speech-accompanying gestures, particularly as they might be considered as evidence of embodied cognition. She is also interested in the functions of gesture for speakers and listeners, as well as sources of individual differences in gesture production.

Steve Jacobs is a doctoral student in the Department of Comparative Human Development at the University of Chicago.

Sotaro Kita is Professor of Psychology of Language in the Department of Psychology at the University of Warwick. His research areas are psycholinguistics and the psychology of communication. The main focus is on gestures that are spontaneously produced during speaking and thinking. The second focus is the development of communicative abilities in children. His research on gesture spans both production and comprehension. It also includes the investigation of patients and neurophysiological methods. He has investigated how gesture and speech production processes are interrelated, what the nature of gesture's self-oriented (as opposed to communicative) functions are, how the listener/viewer integrates information from speech and gesture in comprehension, and how gesture and language develop during infancy and childhood. He has served as the Vice-President and the President of International Society for Gesture Studies.

Susan Levine is the Rebecca Anne Boylan Professor of Education and Society in the Department of Psychology at the University of Chicago. She has an extensive track record of research in the area of mathematical development. Dr. Levine›s research focuses on individual variations in the development of early mathematical thinking, including numerical and spatial aspects of math, and how variations in home and school input affect children›s learning in this domain. Current research projects investigate the particular aspects of adult number talk that are effective in scaffolding numerical thinking. She is also examining individual variations in early spatial thinking and the relation of these variations to exposure to spatial language and engagement in spatial activities such as puzzle play. In addition, she is studying the role of socio-emotional factors in children's math learning, including the math anxiety of parents and teachers as well as children themselves in this learning. She serves as the co-PI of the Spatial Intelligence and Learning Center, an NSF Science of Learning Center.

Tyler Marghetis is a doctoral candidate in the Department of Cognitive Science at the University of California, San Diego. His research addresses the embodied, material, and cultural character of abstract thought, especially mathematics. He has worked on such topics as the embodiment of negative numbers and mental arithmetic, the impact of gesture on the neural response to mathematical equations, the importance of metaphor in the historical development of calculus, and cross-cultural diversity in spatial and numerical cognition. Previously he was a national team wrestler—small wonder he is convinced the body is central to all thought.

Deborah Moore-Russo is Associate Professor in the Department of Learning and Instruction in the Graduate School of Education at the University at Buffalo, State University of New York. Her research interests lie at the intersections of mathematical communication, reasoning, and visualization. Her primary focus is on how individuals understand and engage with fundamental secondary and postsecondary mathematics concepts. More recently, she has started to consider mathematics teachers' reflections and their sense of obligation to the discipline of mathematics. She also researches the use of physical and digital tools to enhance the meaning-making process in postsecondary STEM education.

Professor **Rafael Núñez** is at the Department of Cognitive Science, University of California, San Diego, where he is the director of the Embodied Cognition Lab. He grew up in Chile, pursued his doctoral training in Switzerland, and completed his post-doctoral work at Stanford and Berkeley. He investigates abstraction and conceptual systems from the perspective of the embodied mind. His multidisciplinary approach uses methods such as psycholinguistic experiments, gesture studies, brain imaging, and fieldwork with the Aymara culture in the Andes and the Yupno from the remote mountains of Papua New Guinea. His book, 'Where Mathematics Comes From: How the Embodied Mind Brings Mathematics into Being' (with George Lakoff) presents a new theoretical framework for understanding the human nature of mathematics and its foundations. He is the founding co-director of the Fields Cognitive Science Network for the Empirical Study of Mathematics and How it is Learned, housed at the Fields Research Institute for Mathematical Sciences in Toronto, Canada.

Chris Rasmussen is Professor of Mathematics Education in the Department of Mathematics and Statistics at San Diego State University. His research focuses on the teaching and learning of undergraduate mathematics, with an emphasis on courses that typically function as a transition to advanced mathematics. He has examined how pedagogical approaches and instructional design principles that have been successful at promoting student learning in earlier grade levels can be adapted to the university setting. Central to this work is careful attention to student reasoning and a systematic, theory-driven study of the social interactions in which meanings are established, where norms for convincing arguments and presentations are negotiated, and where students can connect more formal mathematical developments to their personal experience.

Ornella Robutti is Associate Professor in Mathematics Education at the Department of Mathematics of the University of Torino, in Italy. She is a member of several scientific committees in Italy and the European Union devoted to mathematics education, assessment, and lifelong teachers' professional development with technologies. She directs the scientific and agronomic area of pre-service teacher education programme at the University of Torino, as well as the GeoGebra Institute of Torino. She is the author of many publications in the field of mathematics education. The focus of her research is on mathematics teaching and learning as cognitive processes carried out with the support of technologies, and was a member with the University of Torino in the Comenius project EdUmatics, aimed at developing teacher training materials with technologies.

Cristina Sabena is currently Assistant Professor of Mathematics Education in the Department of Philosophy and Education at Università di Torino, Italy. She collaborated in the Italian translation of the book by G. Lakoff and R. Nunez (2000) *Where Mathematics Comes From: How The Embodied Mind Brings Mathematics Into Being.* Her main research interests focus on the role and interplay of different semiotic resources during mathematical activities, with attention to students' problem-solving and argumentation processes. For many years she has also been involved in reflection on the use of different theoretical approaches in mathematics education, contributing to the "networking theory" approach.

Nathaniel J. Smith is a postdoctoral scholar in Informatics at the University of Edinburgh. He has broad interests in the cognitive mechanisms that allow language to be used in real time, and to interact in a fine-grained, flexible, and non-modular way with non-linguistic cognition and action, which he addresses through a variety of methods, including designed experiments and corpus studies of eye-tracking, self-paced reading, EEG, and discourse analysis.

George Sweeney is Assistant Professor of Mathematics at Santa Ana College. His research focuses on how undergraduate students utilize gesture, talk and the environment to do complex mathematics, particularly linear algebra. In addition, he is interested in how members of classroom communities negotiate meaning and develop identities as they do mathematics.

Mike Thomas is a Professor in the Mathematics Department at The University of Auckland, New Zealand. He is interested in semiotic analyses of gestures and representations in learning mathematics, including links to the nature of advanced mathematical thinking, the learning and teaching of calculus, the transition from school to university and undergraduate mathematics, such as linear algebra. He also investigates the role of teacher affective factors, such as orientations and confidence, in school and university teaching and professional development, often with an emphasis on the use of digital technology.

Janine M. Viglietti is Assistant Professor in the Mathematics Department at University at Buffalo, State University of New York. Her research interests lie in the impact of affective dispositions on pedagogical practice. Recently, her work involves studying preservice teacher preparation including integrating assessments of teaching practice. Preliminary work involves crafting assessments involving role-playing scenarios centered on a 'standardized student.' She and her colleagues hope to evaluate how preservice secondary teachers respond to student misconceptions in real time with a goal of improving teacher preparation.

Caroline Yoon is Senior Lecturer of Mathematics Education in the Department of Mathematics at the University of Auckland, New Zealand. She researches how mathematical thinking develops in students as they work on rich conceptual tasks. Her primary interest is in creating new ways of visualizing evidence of students' mathematical thinking as it changes over time. She also studies and designs mathematical tasks that encourage students to visualize mathematical concepts through gestures.

www.ingramcontent.com/pod-product-compliance
Lightning Source LLC
LaVergne TN
LVHW020642300926
849161LV00004B/777